Making Wise the Simple

Dr. Jack Schaap

PRESS

Making Wise The Simple
by Dr. Jack Schaap

Printed in the United States of America

ISBN 1-594675-94-5

Unless otherwise indicated, Bible quotations are taken from the King James Version.

www.xulonpress.com

Table of Contents

CHAPTER 1

Making Wise the Simple

"The proverbs of Solomon the son of David, king of Israel; To know wisdom and instruction; to perceive the words of understanding; To receive the instruction of wisdom, justice, and judgment, and equity; To give subtilty to the simple, to the young man knowledge and discretion. A wise man will hear, and will increase learning; and a man of understanding shall attain unto wise counsels." (Proverbs 1:1-5)

I was 17 when I was called to preach; and almost immediately, I began working with teenagers. I became a youth director and built the youth group from eight to forty-five in one summer. The next summer when I came home from college, I worked with the group again. The doors opened for me to begin accepting outside preaching engagements. Since 1976, I have preached to hundreds of thousands of teenagers. In fact, for many years I preached almost exclusively to teenagers. I did not try to become a teen speaker; for some reason God opened the doors and gave me their hearts. I really wanted to help young people.

From the very beginning of my being in the ministries of First Baptist Church of Hammond, Indiana, I watched my former pastor, Brother Hyles, make many brilliant decisions. Perhaps his most astute decision was to avoid the tragic mistake that King David made, which was serving his own generation. *"For David, after he had served his own generation by the will of God, fell on sleep, and was laid unto his fathers, and saw corruption."* (Acts 13:36) David did build one of the greatest generations in the Bible; unfortunately, he did not prepare the next generation to succeed him. The very next generation faltered horribly, and the following generation was totally abandoned by God.

How important it is for us to do the same as Brother Hyles did—to prepare the next generation for when we are gone. It is not ours to splurge what we have been given.

After I became pastor, someone said to me, "If you've got it, you ought to flaunt it!" Those flippant words hurt me deeply! He was saying, "Man, you're the pastor of that big church! Enjoy it!"

"It isn't mine to enjoy," I said. "I didn't take this position to enjoy it. If it is enjoyable, I must have the wrong definition of enjoy."

I love pastoring First Baptist Church of Hammond, but I do not believe it is mine to indulge.

It is not ours who lead this ministry to indulge ourselves in the ministry. Ours is to prepare the next
generation so that we can keep First Baptist Church
going. According to the Bible, we are always supposed to be living for a generation other than our
own. Seemingly, the mistake that every generation
makes is to live for themselves. I do not want to see
my generation make the same mistake. I believe the
concepts presented in these lessons will help prepare
the next generation.

The Bible describes five different groups of people, and these groups are all interrelated. The first
category and central character for this study, the simple one, is found in Proverbs 1:4, *"To give subtilty to
the **simple,** to the young man knowledge and discretion."* There are many definitions for the simple person. When Brother Hyles comprehensively taught
the book of Proverbs for many months, he gave the
definition for the simple one as "the unpleated one"
or "the uncomplicated one." He said, "The simple
one is the one whose life has not yet entered into the
complications of life or has not yet entered into the
manifold folds, meaning the many folds of life:
mortgage payments, marriage, children, utility bills,
car payments, car insurance, and so forth." I never
forgot his definition.

It's also like the axiom, "As the twig is bent, so
grows the tree." The simple one can be bent toward
wisdom, and then he is always leaning toward wisdom. By nature, the simple one does not want wisdom because he is not intelligent enough to
recognize the value of wisdom.

The Simple One

As my ministry grew, I began to realize more and more that the lives of today's simple ones were becoming more and more complicated. One such complication is more immorality. Fourteen-year-olds and younger children are giving birth to children. I met a lady in East Chicago who said, "Brother Jack, four girls in my daughter's fifth grade class are carrying babies." Fifth grade! Complications have been brought into the zone of life where children and teenagers—simple ones, if you please—are unable to deal with these incredible responsibilities.

A simple person can go in one of two directions, according to the book of Proverbs, to reach an ultimate goal. A simple person can choose to do right and go to the next stage of reaching his goal—prudence. When a simple person chooses the path of the prudent person, he will eventually reach wisdom. The path is simple—simple to prudent to wise. God says to a young person who is pursuing wisdom, "Be careful to choose the right intermediary stage." The wise person is the third category of people addressed in the book of Proverbs. However, if a simple person is not careful, he will make the wrong choice. He will unwisely choose the other direction and meet the scorner, the fourth category of people found in Proverbs. A scorner, if he continues with a scornful mentality, always winds up becoming foolish. He ultimately becomes a fool, which is the fifth group of people. No one is born a fool. He had to go, by choice, through certain stages to become a fool. When he was simple, he chose a scorner as a companion rather than a prudent person.

A simple person must make a choice. He can listen to the prudent person who beckons toward wisdom, or he can listen to the scorner, who will draw the simple one to foolishness. The simple one cannot remain simple because he is in a transitional stage. In the Hebrew, he is a young man—"body of a man, mind of a child." Prior to 1951, the Bible term of *young man* or *young woman* was used. After 1951, a new term was coined—teenager. A teenager is someone who is in transition from childhood to adulthood, and he is learning to make decisions. The main decision is to either listen to the prudent guide and become wise or listen to the scorner and become a fool.

The following are a few more descriptions of a simple one: lacking discretion, silly, open-minded, vulnerable to backsliders and apostates, lacking understanding, gullible, believes every word, inherits folly, lacks discernment, is shortsighted, and is corrected by words and example. A simple person is a blank sheet of paper, and anyone can write on it. Whoever writes first gets to draw the picture.

The Prudent Person

The prudent person protects the simple one's immaturity. He will wisely conceal the shame and hide any embarrassment the simple person feels. The prudent one allows the simple one to learn, to grow, and to mature without making a fool of him. A prudent person can be a parent, a teacher, a bus captain, a pastor. That prudent person guides, protects, nurtures, and nudges the simple along the proper pathway while saying, *"...This is the way; walk ye in it...."* (Isaiah 30:21) A prudent person is one who protects.

A prudent coach understands that his job is not necessarily to win the game; rather, his job is to disciple young men to manhood. The same is true for the umpire, the manager, and the parents at the game. It is very important for every parent to understand that the goal of the youth league program is not to win a competition. It is to have a developing ground for future wisdom and for teaching the youth how to make good choices.

Suppose a boy tries to slide into third base, but he slides too soon and stops 12 feet short of the base and is called out. Maybe a teammate runs home in spite of the fact that the coach says, "Don't run home!" The prudent coach doesn't grab the boy, throw him up against a fence, and shout, "Jerk! You just blew the whole game." No! The prudent coach covers the shame of a young, foolish simple one who has made a mistake. The prudent coach takes him aside and calmly says, "We'll talk about it later." The prudent person gives the child an opportunity to grow, and he nurtures him.

Because God created each person differently, each is handled differently by the prudent person. He finds a way to help the simple get the job done. Prudent parents say, "You have an assignment to do." They guide him, nurture him, and provide the materials needed for him to get the job done.

A prudent person always asks, "How are we going to overcome this problem?" and he gets involved. Prudence is the key to wisdom. The opposite of prudence is not foolishness; the opposite of prudence is scorning.

Simple people are very vulnerable and open-minded. They are unable to discern if the prudent one is right or if the scorner is right. They truly do not know. Consequently, they believe whatever they are told.

Another definition of the simple person is "open-minded" with "no lines or limits." If a scorner gets to the simple one before a prudent man does, the simple one will adopt a scornful life. Children who are constantly berated and called fools usually buy into what it is they are told. Eventually, this person develops a low self-esteem and believes the negativism. However, if a prudent parent gets hold of him and says, "I'm proud of you! You are somebody special! It was a good day when we brought you home from the hospital!"

The child buys into that kind of praise and believes what his prudent parents say.

On the first day of school, a prudent teacher says, "The joy of my life is to be your teacher. I love this job! I'd pay money to work here."

The simple person believes anything—whatever is told to him. Whoever gets to the simple one first— the prudent person or the scorner—sets him on the path he will walk for the rest of his life. The prudent person finds a way to get the job done. He wants the simple one to understand the rules. He therefore works to guide the simple one to reason and understanding. When the simple one becomes wise, he will be able to say to the next generation, "I know what the reason is."

The prudent man learns from the mistakes of others. He has a heart for learning. He has learned to stand back and take a long look at the situations developing around him. *"A prudent man foreseeth the evil, and hideth himself; but the simple pass on, and are punished."* (Proverbs 27:12)

The prudent parent warns the simple one to be careful about dating. Let me illustrate. A young man came to meet with me, and in the course of our appointment, we began discussing his dating behavior. I asked, "Are you hugging and kissing?"

"Of course I am," he said. "I'm 16 years old, and I like hugging and kissing."

"Are you going to marry her?" I asked.

"I don't know," he honestly answered.

I asked, "How would you feel about a boy who hugs and kisses your sister?"

"First of all, it would be gross," he said, "but, if he did more than that, I'd be pretty ticked."

"Okay. Suppose you have a 16-year-old daughter, and some boy wants to be immoral with her."

"I'd kill that rascal," he vowed.

"Do you understand the basic concept that the girl you are dating is someone's daughter and someone's sister?"

He said, "I never thought of it like that before."

That illustration perfectly explains the simple one. The simple one doesn't have the wisdom to make this type of decision, but he will listen to a prudent person who says, "Let's take a long look. Son, how would you feel if you got married ten years from now, and you found the woman you married was kissed and touched by some of the guys in your class?"

"That would be disgusting—especially if So-and-So was the one. I hate that guy," he answered.

"You are getting the picture," I said. "Suppose you marry the girl he is dating?"

"Oh, I would never marry her; she's ugly!"

"But suppose she gets beauty pills," I said, "and suddenly she becomes beautiful."

"I'm with you, Brother Schaap. I would be really angry."

I asked, "Do you realize that if you are not going to marry the girl you are with now, she is going to be someone else's wife. Her husband is going to feel the same way as you do."

"Brother Schaap," he said. "I have never thought of any of this."

That boy was a simple one. Until we met in that appointment, he had never thought of all the ramifications of his actions. Teenagers believe they are old enough to misbehave with a member of the opposite gender, but they are not old enough to think beyond the end of their nose! They have very little common sense.

A simple person might be 16 years old, and all he knows is his girlfriend has beautiful lips, and he wants to put his against hers. The prudent guide says, "Son, before you do that, let's take a longer look at life."

The Scorner

The scorner rejects any kind of correction. He

hates those who punish him; for that matter, he hates those who punish others. If the leader corrects someone, the scorner is the one who gets in the face of his leader. When the prudent coach pulls a scorner to the side and says, "Let me talk to you for a minute," the scorner throws a temper tantrum. If that scorner isn't corrected properly and wisely, he will become a fool. *"A reproof entereth more into a wise man than an hundred stripes into a fool."* (Proverbs 17:10) The Bible says that a look of disapproval to a wise person will suffice; whereas, corporal punishment is meaningless to a fool.

The wise father can look at his son with disapproval, and that boy is crushed because he disappointed a wise man. However, a dad's trying to discipline a foolish son is another matter. Some need harsh scoldings. Some need sound spankings. My mama spanked me 28 days in a row for the same offense. I finally got the message!

On the other hand, the scorner berates, belittles, humiliates, and mocks. He brags about the rules he has broken, he rejects correction, he hates anyone who punishes him, and he tends to be a loner because nobody wants to be with him. He appears to want to do right and is the cause of strife and contention in every institution. Get rid of the scorner, and the strife and contention will end.

The scorner says, "I'll fight you about that. Prove it!" When the umpire says to the scorner, "You're out," the scorner retorts, "No, I'm not!"

The scorner questions every rule and every decision. He doesn't want rules. He doesn't need anyone except himself. The only people who will follow a scorner are fellow scorners, usually because of intimidation and badgering. The truth is that nobody really likes to be with a scorner. In fact, if his peers stood up to him and said, "We don't want what you have," and left him, he would be standing alone. Still, there is a little hope for a scorner. There is no hope for a fool. In fact, the Bible gives the fool one commendation: *"Even a fool, when he holdeth his peace, is counted wise: and he that shutteth his lips is esteemed a man of understanding."* (Proverbs 17:28)

Every child, every teenager is a simple, open, blank page waiting for a prudent guide who understands his job is to lead the simple one to wisdom, or the simple one is waiting for a scorner to defile and poison him and take him to foolishness. The simple one no longer has a choice once he buys in. Simple one, be careful to whom you listen—the prudent guide or the scorner.

CHAPTER II

Who Are the Simple?

"The law of the Lord is perfect, converting the soul: the testimony of the Lord is sure, making wise the simple. The statutes of the Lord are right, rejoicing the heart: the commandment of the Lord is pure, enlightening the eyes. The fear of the Lord is clean, enduring for ever: the judgments of the Lord are true and righteous altogether." (Psalm 19:7-9)

The key word of this chapter is the word simple. The simple one is introduced in Proverbs 1:1-4, *"The proverbs of Solomon the son of David, king of Israel; To know wisdom and instruction; to perceive the words of understanding; To receive the instruction of wisdom, justice, and judgment, and equity; To give subtilty to the simple, to the young man knowledge and discretion."*

The book of Proverbs is based on the simple one arriving at one of two destinations. One of the two destinations is wisdom, which Proverbs calls *"the principal thing,"* or the chiefest of a simple one's choices. The words *"principal thing"* also mean "firstfruits." The Bible is teaching a person is supposed to choose wisdom like God chooses the tithe.

The simple one's other choice is foolishness. A person cannot choose both wisdom and foolishness. They are at opposite ends of the spectrum. A person can exhibit characteristics of wisdom or foolishness, but he cannot be a fool and wise at the same time.

The Bible teaches that the simple one does not arrive at wisdom without the help of a guide. The guide for simplicity to get to wisdom is prudence. The prudent guide takes simplicity by the arm and guides him to wisdom. In the Bible, the word prudence is a term meaning "a wise guide"—someone who knows how to make a path to wisdom.

Likewise, as the simple one cannot get to wisdom without a prudent guide, he cannot get from simplicity to foolishness without another guide, who is called a scorner. The simple one can be assisted to becoming a fool by a scorner. A simple one cannot get there by himself; he will get there only with the help of a scorner.

Keep in mind that the word simple does not describe a little child. A metamorphosis takes place when a child reaches adolescence; he becomes different. No wonder Mark Twain said when a child becomes 13 years of age, he should be locked up in a box and a little hole should be cut in the box. He said when the child turns 18, the hole should be plugged!

The Bible gives several characteristics and definitions of what a simple person is.

1. The simple one is naive. He is a blank sheet of paper. He is wide open, and nothing by design has been written on him. He has no real direction or purpose. The word literally means that he lacks knowledge based on life experiences. He simply has not lived long enough to make any deductions. The simple one is mentally roomy and open-minded. Often, he does not make good judgments and lacks discernment. He has not experienced enough of life to draw any proper conclusions. The simple one is also available. By that I mean he has lots of room for writing on him. He has not had the marks on his life to give him enough direction to make wise choices. The older a person gets, less room is available for anyone to give him direction. On the other hand, the "blank piece of paper" can be written on.

2. The simple one is gullible and vulnerable. *"The simple believeth every word: but the prudent man looketh well to his going."* (Proverbs 14:15) The simple one believes every word!

Romans 16:18, *"For they that are such serve not our Lord Jesus Christ, but their own belly; and by good words and fair speeches deceive the hearts of the simple."* Simple ones are deceived by smooth words. If a person claims to be an expert on a certain matter, the simple one will fall for his line. He is easily enticed. The Hebrew rendering of the words, *"believeth every word,"* is "very easily seduced," "very easily flattered," and "very easily persuaded."

Compliment a young man on his football prowess, and he becomes like putty in another's hands. He easily succumbs to flattery.

On that simple one's blank piece of paper, there are almost no boundaries. They have no restrictions. When a simple one gets into trouble and is questioned about his actions, he often replies, "I don't know." He really does not know! They are not bright enough to know how dumb they are!

Most evolutionists are simple people. They have not been taught the simple truth of Proverbs 1:7, *"The fear of the Lord is the beginning of knowledge...."* Instead, they have bought into what some scorners taught them about the big bang theory, cosmic dust, and theories about the beginning of man. Because the scorners mocked the Bible, the simple person never learned about the fear of the Lord, which is the beginning of wisdom. Without the fear of the Lord, a person cannot begin to learn. He stays naive, vulnerable, and gullible.

3. The simple one is unskillful in war. I carefully referenced over 121 instances of the word *wisdom*. In about 60 places in the Bible, the word *wisdom* refers to the ability to make war. Of course, I am not referring to fistfights or the like; I am referring to spiritual warfare. Because of their vulnerability, simple ones do not possess the knowledge to defend themselves. Ephesians 6:11 says, *"Put on the whole armour of God, that ye may be able to stand against the wiles of the devil."* The simple ones have no idea about the availability of the spiritual armor of the Lord. He must pick up the Sword of the Spirit, read

it, study it, and wield it as a powerful weapon.

Every day I choose a Scripture verse to use as my sword and shield. For instance, one day I chose I Chronicles 28:20, *"And David said to Solomon his son, Be strong and of good courage, and do it: fear not, nor be dismayed: for the Lord God, even my God, will be with thee; he will not fail thee, nor forsake thee, until thou hast finished all the work for the service of the house of the Lord."* That verse tells me not to be afraid of this great work because the Lord is going to be with me from the beginning all the way until I finish the work He has called me to do! Every time I take out that 3x5 card and read that verse, I am taking out a piece of armor to fight the enemies of doubt, worry, and fear that come my way. Simple people have not yet realized how important it is to put on the armor of God.

I am often asked, "Are only teenagers simple people?"

No! I know many simple people who are my age, which is 46 at this writing. I also know some who are 26, 36, 56, and 66. Simple is a character; simple is not an age. Many older simple people are blank— still gullible, still vulnerable, still naive, and still unskillful in war. Since they do not know how to fight temptation, they succumb to it.

Simple people do not know how to do battle. They do not wisely use the weapons of warfare like the shield of faith. They simply do not know how to use the promises of God to protect them. They do not know how to put on the helmet of salvation, nor do they have their feet shod with the preparation of the

Gospel of peace. Their loins are not girded about with truth. As a result, they are vulnerable to attack and instead of doing battle, they just take it. They do not know how to fight back.

4. The simple one lacks sound decision-making principles. In fact, the simple one does not know how to make a decision. The job of parents is not to make decisions for their teenagers. One of the greatest mistakes a parent can make is to make all of their teenagers' decisions for them. I believe making a teenager's decisions is just as big a mistake as giving kids all the money they want without making them earn it. What I call "a welfare mentality" is created in the children. An incredible amount of damage is wreaked on children spiritually, morally, emotionally, and mentally when they are not taught how to make decisions. I personally feel the biggest role I play as a father for my teenage son is to guide him into making proper decisions. The decisions the simple ones make are at best 50/50.

For instance, I never chose our son-in-law, Todd Weber, as a husband for our daughter Jaclynn. Rather, she decided for herself, and I approved her decision-making process. Many times she came to me with decisions she had to make and would say, "Dad, I have to make a decision about thus-and-so. What should I do?"

"I can't tell you," I would say. "One of these days I will not be here to make these decisions for you, so I cannot make it for you now. My job is to help you know how to make decisions." Parents, stop making decisions for your teenagers and young adults.

Since I became pastor of First Baptist Church, I have counseled with many people. Some people come to see me for advice; at times, I am confronted with people with poor decision-making principles. I am thinking of one young man who came to see me and let me know that he disagreed with me on some issues—basically dating issues.

I asked, "Are you a college student?"

"Yes, I am," he answered.

"What year in school are you?"

"I have been here for three months; I am a freshman," he replied.

"What is your question?"

"First, I really don't approve of the way your college promotes dating," he announced. "I believe dating is unscriptural."

"Okay," I said. "Are you dating?"

"No!"

I asked, "Have you ever dated?"

"No!"

"I'm married," I started. "Who knows more about dating, you or me?"

"I do!" he exclaimed.

"I was afraid you would say that," I said. "Since you have so much wisdom, why don't you tell me what I am supposed to do about the dating problems."

"Well," he said, "I think your teachers are pushing dating—guys dating girls and girls dating guys."

"Would you like to see guys dating guys and girls dating girls?"

"No! I just don't believe guys should be dating girls, and girls should be dating guys. I believe in betrothal. I believe the parents should choose their child's future mate."

I asked, "Can you give me a scriptural background for believing in betrothal?"

When he really couldn't give me an example in the Bible of a parent choosing a spouse for his child, he sputtered, "Abraham."

After he argued the point, I explained that Abraham did not choose a bride for Isaac; rather, he sent his servant for Rebekah. I further explained that the Bible says, *"...a prudent wife is from the Lord."* (Proverbs 19:14) What a classic example of a simple one!

Many young people come to my office to ask questions like:

- "Should I marry So-and-so?"
- "Should I be a preacher?"
- "Should I go into evangelism?"
- "Should I pastor such-and-such a church?"

I cannot answer these questions. I do not know! That is your decision! I can tell a person how to make that decision, but I will not make the decision.

I went through the decision-making process when I graduated from Hyles-Anderson College. I had to make a choice when Brother Hyles asked me to teach English at the college. I agreed to teach for five years. When five years was up, he had an appointment with me. He told me to give him an answer in two weeks about becoming an administrator.

Because of those meetings with Brother Hyles, I

wrote down 12 principles of how I would know God was leading me. Those principles helped me make decisions in my life. In the same way, my wife and I devised seven rules for spanking our children. We did not spank them unless the infraction had to do with breaking one of the seven rules.

The truth of the matter is, a wise person knows how to make decisions. In all actuality, a wise person doesn't have to make many decisions. He has learned that principles make the decisions for him.

Wisdom says, "My principles make my choices for me. Since they make my choices for me, I really do not have choices to make." Following principles helps a person to live line upon line, and precept upon precept. The prudent one teaches the simple one how to make decisions. Too many people want others to rubberstamp their decision-making.

A young lady may ask, "Should I date So-and-So?"

My first question is always, "Is he saved?"

Perhaps the questioner responds, "No," but quickly adds, "I think I can get him saved."

It has become a moot issue. Principle number one has been violated! My Bible says in II Corinthians 6:14, *"Be ye not unequally yoked together with unbelievers...."*

A young man came to see me and announced, "I'm in love with So-and-So. Should I marry her?"

"What are you going to do with your life?" I asked. When the person did not know, I said, "You cannot marry anyone because the goal of your wife is to be a helpmeet for you. A helpmeet is to help you succeed at what God has called you to do. How can she possibly help you if you do not know what you are supposed to do?"

He exclaimed, "I love her!"

When I told him he did not know what love was, he proceeded to tell me that he loved her because she was pretty and he had great feelings for her.

"None of that is love," I explained. "If your wife is supposed to help you succeed at your calling in life, how can she ever do that? You are going to frustrate your wife, and eventually you will start having disagreements. You are setting yourself up for divorce or great dissatisfaction because your wife's job is to help you succeed at whatever God's calling is for you."

He is a simple one! Proverbs 24:27 says, *"Prepare thy work without, and make it fit for thyself in the field; and afterwards build thine house."* Establish where you are going, get your preparation to that end, and then add a wife. However, most people do not follow a decision-making process. They just blindly let others make the decisions for them. They want someone to simply wave a magic wand and make all the frustration go away. Those who do wave the magic wand are scorners and fools.

5. The simple one lacks practical understanding. They lack sound leadership. They possess very little, if any, common sense. In addition, they have no practical skills. Simple ones have no practicalness, nor are they sharp or clever in practical matters. Does your teenager know how to balance a checkbook, or how to withdraw money from the

bank, or how to talk to the banker? Does your son know how to change a flat tire? Can your daughter prepare a simple meal? Do your teenagers know how to do the practical things in life?

Certainly every Christian young person should memorize Bible verses and be an excellent soul winner, but every red-blooded American male would rather be married to a woman who knows how to keep the house clean or how to cook a good meal than to one who only knows how to go soul winning. Please don't misunderstand me! I do not want non-soul winners to hide behind this illustration, but too many simple ones lack practical understanding.

In fact, one thing I love about my wife is that she not only memorizes Scriptures and goes soul winning, she is also an excellent cook. She has several cookbooks and files she uses. After our daughter married Todd, she was visiting, and I enjoyed watching her go through her mother's files to pick out recipes. Several times I heard her say, "Dad likes this; Todd will probably like it, too."

Not long after, they invited us to their apartment for a meal on a Sunday afternoon. The meal was wonderful, but the great pride I felt was the way in which it was laid out so professionally and so wonderfully. I said to myself, "Thank God for a wife who taught my daughter practical things."

Sometimes I think we are teaching all of the wrong subjects. I do not mind your learning algebra, trigonometry, or social economics. In all honesty though, not one of those subjects will help a couple have a happy marriage. Proverbs 31 tells us that a hardworking woman who strengthens her arms makes a happy marriage. A wife who helps wisely with the finances and contributes to the household makes a happy marriage. Her husband can safely trust her. Those are ingredients to making a happy marriage.

I exhort you parents to train your children and teenagers in the art of making decisions and performing practical skills. Some of the biggest decisions ever made in life are made during the junior high and high school years. Simple ones need a lot of direction.

6. The simple one lacks discretion. Proverbs 14:15 says, *"The simple believeth every word...."* They do not have the ability to recognize the different masks of wrong which come in a variety of choices. The simple one can be taught one kind of wrong, and he can grasp that. Put the same wrong in a different mask, and it will fool him every time.

Children are often better at discerning wrong than the simple ones. For instance, during Vacation Bible School, I had a "special teacher" as a guest for the theme of "Back to School." Ms. Potter walked to her school desk, noticed a Bible, and said, "We don't need this," and dropped it into the trash can. She then proceeded to cut up the flag because it was pretty. The kids watching this so-called teacher nearly went ballistic! They caught the wrong from the very first of the program.

Let's take the example of music. The majority of Christians know and believe that loud music which is all rhythm and beat is bad. However, if that same music has the rhythm slowed, the volume turned down, and is called country and western or rock-a-billy, somehow it has become good. As a matter of fact, it is the same garbage wearing a different mask. Simple people cannot discern the changing masks of evil or wrong.

Brother Hyles used to teach, "The great curse of Christianity is the inability to take a principle and apply it to a multitude of situations." Some people have to be told every time, "That music is wrong!"

Many try to excuse their so-called Christian music with statements like, "The words are Christian." If music has a worldly beat and a hot rhythm, it's still wrong. It does not matter if the song is "Jesus Saves" or "The Devil Saves."

7. The simple one possesses curiosity without discernment. Proverbs 7:7 addresses the young man who is *"void of understanding."* A wise man watches this young man get into moral trouble. *"For at the window of my house I looked through my casement, And beheld among the simple ones, I discerned among the youths, a young man void of understanding."* (Proverbs 7:6, 7) The wise man happened to be looking out of his window and noticed a young man walking toward the wrong part of town. The term "young man" means "young of mind, adult of body." This does not mean stupid. It simply means he was not using discernment; he was not thinking clearly.

Proverbs 7:8 continues, *"Passing through the street near her corner; and he went the way to her house."* He was going to check out the situation. He is like the young man of today who says, "I just walked by the adult bookstore to see what it looks like from the outside. I'm not going inside." Or he might justify, "I'm just going to walk by the magazine rack. I'm not going to read a dirty magazine; I'm just going to look at the cover pictures." Perhaps he sees a billboard about a gentleman's club. "I wonder what it's like," he thinks. The simple one just wants to check out the wrong; he is not looking for trouble.

"I'm not going to smoke cigarettes. I just want to see what they look like and what they smell like. I'm not going to drink beer. I just want to smell it or take one small taste. I don't want to date the girls at the beach. I just want to see what it's like at the beach. I want to cruise the roads along the lake."

The problem with the simple one is that he doesn't think. All he wants to do is hang around and check out the situation. He has no intention of getting into trouble.

My parents had a very strong hatred for cruising and hanging around. In my hometown, the local punks hung out at a corner. My mom and dad would say, "Look at them! That's a bunch of hoods—just hanging around!"

"Boy," I thought, "there's something bad about hanging around!"

The young man in Proverbs 7 was hanging around in the wrong place. He had no intention of getting into trouble, but he lacked discernment. That curiosity and lack of discernment inevitably gets a simple one into trouble. Proverbs 7:23-27 continues, *"Till a dart strike through his liver; as a bird hasteth to the snare, and knoweth not that it is for his life. Hearken unto me now therefore, O ye children, and attend to the words of my mouth. Let not thine heart decline to her ways, go not astray in her paths. For she hath cast down many wounded: yea, many strong men have been slain by her. Her house is the way to hell, going down to the chambers of death."* I have no doubt that boy had no intention of going to Hell. He gave no thought to the possibility of catching a communicable disease. He was just curious, and he did not have any discernment. One of the distinct characteristics of a simple person is curiosity without discernment or discretion.

8. The simple one does not have the ability to discern between many options. Proverbs 9:4-6 says, *"Whoso is simple, let him turn in hither: as for him that wanteth understanding, she saith to him, Come, eat of my bread, and drink of the wine which I have mingled. Forsake the foolish, and live; and go in the way of understanding."* One of the brilliant marketing devices of the Devil is the multiple translations of the Bible. The Devil does not have one choice. The Devil wants any choice but God's choice, and one way to confuse simple people is to give them too many options. To keep the simple one on the way to wisdom, the prudent man offers few options, and he will make the options very right or very wrong.

Some may say, "You fundamentalist preachers are so graphic when you preach against sin. You make it so blunt to the teenagers."

The job of the preacher is not to confuse the

teenagers. We must keep in mind they are simple. The job of the preacher is to make it very clear to them exactly what is right and what is wrong. I don't believe any teenager has ever walked away after hearing one of my sermons on rock music and asked another, "So, what does he think about rock music anyway?" I am very complimented when someone accuses me of being too blunt with the teenagers. My goal is to make it abundantly clear because a simple person does not need shades of gray. They need black and white, on and off, hot and cold, in and out, and nothing in the middle! Simple people cannot discern between the options. The word *discern* actually means "to discern between the shades of color."

Some say, "Well, you know, some things aren't really black and white!"

Don't teach that kind of philosophy to teenagers! Make it black and white for them because they need a lot of clear understanding. By the way, all of God's people need the same! We do not need multiple gray areas in our lives.

For instance, I believe that the King James Bible is the only perfectly preserved English text for English-speaking people. If I did not believe that theologically, I would still choose one Bible. As a prudent man trying to help children grow into adulthood, as a father who wants my son to become a responsible man and my daughter to become a responsible woman, I would choose one Bible. You will not bring anyone to maturity by saying, "Here are multiple choices of authorities."

If I had not liked Eddie Lapina, who was our

children's youth director, I would still stand with him. If I had not liked Dr. Tom Vogel, who was our children's junior high principal, I would still stand with him. Why? It is so important that parents stand with a singular authority in their children's lives. Help your children learn discernment by choosing the right authority and then standing with that authority. Simple people do not see the wisdom of choosing a singular authority.

It is dangerous to say to simple ones, "Well, the principal is one person who has his own opinion, and his opinion is not always right." It is just as a dangerous to say, "The King James Bible is one of many versions that are good." Scorners like to magnify the differences in the leaders. They like to magnify how one leader sees it differently from another. Scorners make comments like, "Now that was a good sermon, but let me give you my spin on the subject." They mock the authority.

Choose the right church and stay with that church. Why? It will help with the generations which follow. Children will turn out much better percentage wise if parents will say, "This is our church." Staying will build security for the simple ones.

Parents should agree with each other when they do not agree with each other. It is so very important because a simple person cannot discern the separate shades of gray. Provide one authority for the simple ones to follow at home.

I have never once said, "You know, what Brother Hyles and I believe is similar, but we disagree on a few matters." I suppose no two people think exactly alike, but you will never know if I have ever disagreed with Brother Hyles. In leadership, it is important for the leaders to be on the same page. As long as I have a sane mind, you will never hear me say, "For years Brother Hyles taught thus-and-so, but let me straighten you out."

My job is not to straighten out another leader's opinion. My job is to either endorse it or shut up. Simple-minded people cannot discern between the subtle shades of gray.

I once interviewed someone for a position; in fact, I had offered him a job. We were just talking, and the person said, "To be honest with you, this is the greatest Bible college in all the world, but everyone has differences. I believe it is important that people you bring on board are open-minded. I am probably 90 or 95 percent just like you. "

"Why are you different?"

"Because I believe it is important that we have some differences," he said. "That difference maintains a unique flavor in all the different departments."

Needless to say, that man is not employed here. Leadership has to be on the same page. Leadership must make wrong exceedingly wrong and right exceedingly right. Subtle shading is what brought sin into the Garden of Eden. Eve was a simple lady who had not matured, and she fell for the Devil's lie just like everyone who is simple-minded. Eve lacked discretion.

9. The simple one is shortsighted. *"A prudent man foreseeth the evil, and hideth himself: but the*

simple pass on, and are punished." (Proverbs 22:3) That very same verse is repeated verbatim in Proverbs 27:12. All verses are important, but if God repeats a verse twice, it must be extra important. I believe God repeats this verse twice because He wanted us to see that the simple one is very short-sighted. Simple ones desire immediate gratification. They live for this moment only. One of the great cardinal characteristics of a simple person is his inability to take the long look and see where his decision will take him.

The simple follow immediate pleasure. For teen-agers it is, "What are we doing today? What thrill can I have right now? How can we have fun?" Teen-agers will wait in line for an hour for a three-minute thrill.

Many important things in the Christian life become fun that do not start out as fun. Let me name a few: duty, work, obedience, integrity, keeping your word, keeping your testimony, guarding your reputation, and preserving your honor. I have never heard of having an integrity party, but people of integrity keep their word, stay married, and build strong homes. People of integrity do not declare bankruptcy. They pay their bills on time. They do not get fired because they work hard, and they are loyal to their employer. Integrity is fun over the long haul.

One area in which the simple are shortsighted is in the area of music. In the 1960's, Elvis came on the scene. He sang good, godly, Gospel songs with just a hint of rock 'n' roll. Then the Beatles came from Liverpool, England. The 1970's worldly music brought hard rock. Woodstock was more than just a rock party. The 1980's brought shock rock and bazaar rock. With the 1990's came alternative music and Gothic music. The Columbine killers were listening to Gothic music which is nothing more than murder music. Has music really gotten fun and exciting? Is killing people fun and exciting? Is murdering your classmates at school fun and exciting? Is being put in jail for life fun and exciting? Since the 1960's, music has made a radical departure from what is acceptable. However, the simple one is shortsighted and cannot see the end result.

The wise man says, "I see a problem."

The simple one says, "It's just music."

Wisdom is not trying to take anything fun from the lives of the simple ones. Rather, wisdom is trying to provide them with real joy and real happiness, and a long-lasting beautiful life.

CHAPTER III

The Need for a Watchman

"I love the Lord, because he hath heard my voice and my supplications. Because he hath inclined his ear unto me, therefore will I call upon him as long as I live. The sorrows of death compassed me, and the pains of hell gat hold upon me: I found trouble and sorrow. Then called I upon the name of the Lord; O Lord, I beseech thee, deliver my soul. Gracious is the Lord, and righteous; yea, our God is merciful. The Lord preserveth the simple: I was brought low, and he helped me." (Psalm 116:1-6)

One Hebrew meaning of the word *preserveth* in Psalm 116:6 is, "The Lord is a watchman to me." A watchman is one who leans forward and peers into the distance. A watchman is someone who spies, or observes, or keeps watch. It is someone who coaches or guards. To help him toward wisdom, the simple person needs what the Bible calls a watchman. The prudent man, who is a watchman, is leaning forward and peering into the distance. He is looking for the path that the simple person has to tread. He is looking for the pitfalls along life's path.

A watchman is like a lifeguard. That lifeguard does not try to keep people from swimming; rather, he tries to provide a safe place for people to swim. The lifeguard watches the swimmers from a high observation tower so nothing will obstruct his view. Markers are placed in the water to provide off-limit zones. When someone crosses the line, he instructs the swimmer to return to the safe zone. He is not trying to stop a person from swimming; he is trying to provide a safety zone. He is not trying to take something from the swimmers; he is trying to provide a place of enjoyment. The word lifeguard is an excellent description of a watchman.

Another good example is the driver's education instructor. The driver's education teacher sits in the right front passenger's seat beside a teenager who has never been behind the wheel of a vehicle. That instructor is not trying to stop the student from driving; rather, he is trying to give him what he wants— his driver's license. The instructor is peering ahead, watching for the traffic signals and the other drivers. He is the one who is driving defensively while the student is learning how to drive. This is another excellent example of a watchman. The driver's education instructor—the watchman, if you please—is providing a safe way for a student to participate in that which he should be doing.

Some leaders become policemen with radar guns. They hide behind corners, lying in wait for the simple one to err, and says, "Caught you. I have my demerit pad." Some parents hide behind a corner, waiting for their child to do something wrong that warrants punishment. Sometimes we have a mentality that we want to catch people doing wrong as though we earn "brownie points" for how many bad deeds we catch them doing.

While I served at Hyles-Anderson College, more than once I heard the president, Dr. Wendell Evans, lecture about the two kinds of people who work at the college. He said, "There are rule people, and there are people people. The rule people cannot wait to catch students doing wrong, and the people people look the other way and want to get along with everyone. These two positions are the extremes on the wrong side."

I do not want any leader to go on a witch hunt, looking for someone to do something wrong. Neither do I want someone to be liked so much that he looks the other way. A certain psyche exists in some people who somehow feel more secure if they can catch someone doing wrong. Somehow that discovery bolsters their righteousness and makes them more holy.

A watchman is not looking for someone to do something wrong so he can catch him and correct him. No! The watchman is looking to keep an individual from doing wrong. The desire of the watchman is to keep a potential wrongdoer from becoming a wrongdoer.

I am in favor of catching people doing wrong. For instance, I believe a parent has permission—a legal, moral, and godly right—to inspect and check all of his children's belongings. A parent has a right to look through his children's dresser drawers, look inside his closet, and check his school bag. I do not think parents should conduct searches to find wrong-doing; rather, I think parents should check to keep wrong from getting into their children's lives. The parents need to have the right attitude and motive. The Lord says, "I can help a simple person, but I am not on a witch hunt to catch someone doing wrong because I cannot wait to punish him."

When I was in college, a very wise man carefully monitored his role in my life. He kept track of my wrongdoing, and like a lifeguard, he kept me within the proper boundaries. He let me continue to swim, but he did not pull me from the pool. He may have salvaged my future ministry because of the way he guided me in college. He helped me walk circum-spectly. He was a very prudent man who saw that if he handled me properly, he would receive total com-pliance from me. God says the simple one needs a watchman.

Many complain about the public school teachers going on strike. We say, "They ought to get merit pay." What if we put parents and Christian school teachers on a merit pay system? Teachers at Hyles-Anderson College and the Hammond Baptist Schools are not paid on the basis of how many demerits they assign. No! The goal of the teachers is to help the students; it is not to harm them or punish them. The purpose of the schools is not to kick out all of the "bad" kids; rather, the purpose of the schools is to train simple people to wisdom. It is to preserve the young people from trouble, hopefully without taking them out of the place where they are learning.

The simple person needs a watchman because of insecurity. He is insecure, and often that insecurity is a result of not developing confidence as a teenager. Young people who do not reconcile past problems do not learn the path of wisdom. They carry the baggage of unresolved problems with them throughout their adult lives.

The simple one is vulnerable to emotional and spiritual oppression or distress. In Psalm 116:6, the Bible says, *"...I was brought low...."* Simple ones are easily intimidated and embarrassed. Because they are vulnerable to insecurity, they seek strength wherever it is. Many simple ones join gangs. They are often caught up in what is coined as "mob psychology." Most who join gangs do so because of intimidation. They prefer not to be in a gang, but belonging to a gang is the only "safe" place to be. They did not join the gang because they wanted to be a member; they joined because they would have been in trouble if they had not joined. The simple ones naturally choose the very source of insecurity to fight their insecurity. The simple one is scared, easily intimidated, easily put down, and easily embarrassed. Yet, simple people turn to the very cause of their vulnerability and seek strength from the cause!

Not only do I see gang members seeking security from that which causes their insecurity, but I also see the same dependence in abused young people. Perhaps a girl has been abused by a relative or has been taken advantage of by an older person. Guess where those girls go for security? They turn to illicit behavior! They do not turn to righteousness. They turn to the very sin that gave them the insecurity to find security. I have yet to find an exception to that rule in over 27 years of preaching. Of course, I am not addressing a girl who does a singular act of misbehaving with a boyfriend; rather, these girls have a record of misbehavior morally.

I counseled a college-age young man who was attending a state college. He informed me that he did not believe in that "religious junk" which was merely a crutch for weak people to lean on.

I asked him, "May I ask you a question?"

"Sure, you can," he replied.

"What terrible thing happened to you when you were a teenager or a young child?"

He visibly reacted to my statement—his face paled, and big tears welled up in his eyes. "Who told you?"

Of course, no one told me. Because I counsel so many teenagers, I see patterns in their lives. I talked at length with that bitter young man. He told me about a horrific incident that happened while he attended a private school. He also confessed to becoming a part of the same horrific deeds perpetrated on others. Then he said, "I hate everything about me. Why would I do the very thing that caused me great harm?"

"Because that is what insecure people do," I truthfully answered.

Sad to say, simple people turn to the very sin that gave the insecurity as a way to combat their insecurity. They only get deeper into their sin as they keep on perpetuating their insecurity. What a vicious circle!

I was on Rush Street in Chicago, passing out tracts when I met a 51-year-old man, who would not take a Gospel tract. When I asked him what happened to him when he was a defenseless child, he told me that as a seven-year-old boy, he was molested by a man in his church.

A watchman must be very aware of the fact that when a teenager gets into trouble, more than likely he will go deeper into that sin. Adults like to think that they will never do the same thing again if they are caught, but they do! They do it again and again and again and become an expert. That is why it is so important, when a child or teenager does get into trouble, that he is handled wisely. Take the simple one to someone who is wise and can help him work through his problem. One of the worst ways to handle trouble in a young person's life is to cover it. The situation cannot be overlooked by saying, "We don't want to tell anyone what happened."

A child who fornicates needs to be confronted by the leaders because if the situation is not dealt with, the fornication will continue. Those who are guiding the simple ones should say, "Let's get to the pastor.

Get to the Christian school principal. Let's see the youth director." What a simple one does not realize is that when he commits a vile sin or has one committed against his person, it brings an emotional baggage that he cannot handle or carry.

I was reading an article in *Time* magazine about bipolar disease, formerly termed manic depressive disease. I was surprised to learn (as are the doctors and scientists) that since the mid-80's, symptoms of bipolar disease have begun to appear in young people as early as 12 years of age. Before the mid-eighties, this condition, characterized by extreme mood swings, was rarely diagnosed in an individual under the age of 30. The author of this article wrote that the only explanation the scientific world can find for this phenomenon is the young person's dealing with the high stress of living with their parents' divorce.

In this day and time, simple people are exposed to vulnerable insecurities before they are old enough to deal with them. In fact, instead of dealing with them, they plunge deeper into their insecure world through the tools of Gothic music, drugs, and sexual promiscuity

Scientists say that the brain doesn't even finish its connection until a child reaches the age of 15. Scientists recently discovered that the logic gap between the left and right hemispheres of the brain makes its final connection in a person at the age of 15. The brain does not stop growing until the person reaches 30. This discovery alone validates why 14 year olds are so illogical. That is why it is difficult for a prudent person to capture a simple one and build his trust.

The Bible says in Psalm 116:6, *"The Lord preserveth the simple...."* The word *preserveth* means "to hedge in, to guard, to protect, to attend to, to keep within bounds, to refrain, and to abstain." I personally believe one of the biggest mistakes parents make is by putting too much trust in their teenagers. Simple people cannot be trusted. Don't get me wrong. You can give them some money and trust them to go to the store and buy a certain item. I am not addressing that kind of trust. I am addressing the kind of trust parents place in them that says, "Well, they're home, and they won't do anything bad."

Parents, do not leave your young people home alone hour upon hour and assume they won't do anything wrong. Too many people say, "You don't know my children; they won't do anything bad. They promised me."

Oh, yes, they will! I know human nature. I don't even trust me! If I know me and don't trust me, how can parents expect me to trust their children? Far too many children get into trouble because Mom and Dad believe their simple ones are worthy of their trust.

Don't leave a teenager home alone with cable television, access to HBO, Showtime, or other movie channels. Don't leave a teenager home alone with access to the Internet. If you want cable television in your home, wait until your children are adults. Then monitor each other.

Jeremiah 17:9 says, *"The heart is deceitful above all things, and desperately wicked...."* Never allow your teenager to be with friends without supervision.

That is what is meant by "preserving." Walk guard around your simple ones. Walk in on them.

If your daughter asks, "Mother, what are you doing?" you answer, "Oh, I'm just playing watchman." Close the door and check in on them again.

The truth is, if you enjoy being with your children and if you are with them much of the time, they won't get into trouble. The children who end up in my office with sad tragedies are, for the most part, children who have godly, sincere, caring parents who are shocked by the whole incident and say, "I cannot believe my child did that." I can believe any young person could because we are all made of the same flesh, and young people are simple.

When Jaclynn was 13 years old, we gave her a promise ring. The ring was a symbol of a commitment from her to us that she would stay pure until her wedding day, but that ring was not her guard; her mother and I were her guard.

When Jaclynn was just a few weeks from marriage, my wife said to her, "You are madly in love. Now there is no human reason why you should remain pure because you are almost married. Now is the time that you have to be really careful."

I say to all engaged couples, "Walk very circumspectly during your engagement." I say to parents, "Watch them better than you ever have. Don't just say, 'They're good kids.'"

Good kids need a watchman to walk guard constantly and watch them all the way to the marriage altar, or they will get into trouble! They are, after all, still simple ones.

CHAPTER IV

Prudent Scorners or Foolish Guides

"Doth not wisdom cry? and understanding put forth her voice? She standeth in the top of high places, by the way in the places of the paths. She crieth at the gates, at the entry of the city, at the coming in at the doors. Unto you, O men, I call; and my voice is to the sons of man. O ye simple, understand wisdom: and, ye fools, be ye of an understanding heart." (Proverbs 8:1-5)

I have carefully studied every verse in the Hebrew that talks about the words simple, prudent, scorner, and wise. I also studied every verse in the Greek that contains those same words. As I studied the scores and scores of verses, I kept coming up with the same definitions for simple again and again and again. In my mind, it seems God was trying to reinforce what the simple ones are and what the bottom line characteristics are.

The three definitions that kept appearing about the simple one were "silly," "seducible," and "unsuspecting." Allow me to explain each of these characteristics in detail.

1. Silly. Many of the words I looked up said a simple person is a silly person. When we say, "Oh, that's a silly joke," or "He's a silly person," that is not the Bible definition of silly. The word silly means, "having or showing little sense," or "foolish, absurd, ludicrous, or irrational." A silly person does some absolutely ludicrous, absurd things. He does foolish, irrational things. Proverbs 8:5 says, *"O ye simple, understand wisdom: and, ye fools, be ye of an understanding heart."* According to this verse, the other group of people associated with the simple ones are fools. A simple person has much more of a gravitational pull or a magnetic attraction toward foolishness than wisdom. To a simple person, wisdom is very unnatural.

Primarily, the book of Proverbs refers to teenagers because in the natural state of progression, a simple person is a teenager who has not experienced enough of life to make proper decisions. To be sure, he needs someone wiser to guide him. A simple person is further from wisdom than foolishness. A simple one left to himself will automatically turn toward foolishness. It is much more difficult for a prudent man to guide a simple one to wisdom than it is for a scorner to take a simple man to foolishness because simple ones have a natural tendency to bend toward foolishness.

As I have already established, *simple* means "having or showing little sense." Simple also implies irrational behavior that demonstrates a lack of common sense, good judgment, or seriousness. In other words, the simple person rarely takes anything seriously. How many times does the teenage crowd have to be publicly scolded? The leader often has to make statements like, "Take this seriously," or "Stop your cutting up." The simple ones have to be scolded more than any other crowd because they tend to take matters less seriously. *Sober* is the Bible word which means, "taking life seriously or taking matters seriously."

Though the simple one is characterized by irrational behavior, silly does not imply someone who is mentally lacking. For instance, some very good people came with their son to counsel with me. After I heard the story, I looked at the boy and said, "Son, you've got stupid disease, don't you?"

The boy sheepishly chuckled and said, "Yes, sir, Brother Schaap, I've got stupid disease."

When I say, "stupid disease," I do not mean "stupid" as in "mentally slow." Silly does not mean stupid; rather, silly means lacking the power to reason. In other words, a simple person by himself does not have the reasons why he does what he does. He has emotions for doing what he does, and he is easily persuaded.

Silly does not imply being unreasonable. It means, "he does not have any reason." An unreasonable person is someone who is willfully prejudiced. He is apt to say, "I don't care what you say, this is what I believe." Actually, that is an accurate description of most adults. For instance, if you see a teenager while soul winning, he will generally listen to the plan of salvation. It is amazing how quickly a teenager will trust Christ as his Saviour. Ordinarily, he doesn't listen to all of the reasoning by the soul winner, but if he does, it quickly makes sense to him.

An unreasonable person is someone who says, "I don't care what you tell me!" That kind of statement is a prejudicial willfulness that says, "I'm right even if you show me otherwise."

2. Seducible. The word *seducible* is derived from the word "seduce," and it means "easily misled." By nature, a simple person is very easily misdirected. He just naturally goes astray. Left to himself to make his own choices or his own decisions, he will naturally and automatically drift toward foolishness—just like a flower bends toward the sun. Turn the flower pot away from the sunshine, and the flowers will bend back toward the sun. The simple automatically crave foolishness. Actually, the simple and the fool are very alike. In fact, the simple one and the fool can stand side by side, and many times you cannot tell the difference. Only a prudent person and a wise person can discern the difference.

Prudent people have to be careful because the simple one and a fool look very, very similar. The difference is that a fool is a fool by choice; a simple one is simple by ignorance. The simple one acts foolishly because he does not know how to take life seriously. The simple one should be treated more delicately than a fool.

Through the years, I have watched many youth speakers abuse the crowd. Since they do not know how to teach or reach a young person, they resort to lashing out at them—yelling at them, berating them, and humiliating them. Certainly there is a time to correct publicly or talk frankly with them. I am not against that, but I am against the berating.

Sometimes a prudent person makes a mistake in not recognizing the simple one. Because of lack of recognition, the simple one is treated like the fool should be treated. That lack of awareness drives the simple one further toward making the choice to become like the fool.

Seducible also means "easily persuaded"—especially to be disloyal or disobedient. When I was growing up, I watched neighborhood boys gather by the railroad tracks and sneak a few beers or smoke a few cigarettes. Some would sneak a dirty magazine to school and say to anyone who would listen, "Want to come see it?" Those boys who took a look gave no thought of what would happen if their parents or teachers caught them. They were very easily persuaded. It is easy to get a simple person to do wrong—not just because of the sin nature; it's just the nature of a simple person. He is one who is very easily talked into being disloyal.

Probably 15 years ago on a Wednesday night, I was standing with my wife in the hallway by Brother Hyles' office. I noticed a lady stop Brother Hyles. I recognized her as being well-known in our church at that time. I felt she and her husband and her family were inner-core members of the church. Brother

Hyles asked her what he could do for her.

"This is our last service," she answered. "We wanted to come and say goodbye to you."

"Are you moving?" he asked.

"No," she said. "We are leaving this church." I could tell by the shocked look on his face that it was like someone had kicked him in the stomach. Big tears began running down his face, and he said, "I'll call you tomorrow."

About a week or two later, I asked him, "Did you call So-and-so?"

When he acknowledged that he had, I asked, "What happened?"

"Jack," he said, "I just cannot figure out people. If you would have asked me to name the top ten families whom I thought were grounded in this church, they would have been on the list. When I called her the next day and asked why they were leaving, she said they did not agree with a decision I had made. That decision had absolutely nothing to do with the church."

Brother Hyles went on to tell me that she disagreed with a personnel decision. She did not agree with his decision to hire a certain person for a position in one of the schools. The family had been in the church for 30 years, and by her word, they had never before disagreed with Brother Hyles on any matter —not one! They chose to leave over their very first disagreement! I call her a simple woman.

Wise people do not jump ship on what they think is one bad decision. Wise people say, "Till death do us part" with their marriage. Wise people say, "I brought you into this world, I'm staying with you."

A worried parent told me that his 17-year-old son wanted to leave home. He came to see me, wondering what to do.

I said, "Tell him to stay home."

Instead, the father let him go. I do not understand that decision. Until my children are adults, I am going to do everything in my power to keep them home until it is time for them to leave. All of their lives, I am going to stay with them in my support. If they break my heart, I am still going to stay by them.

I told Jaclynn before she was dating her husband Todd, "If you run away from home, live in sin, and come riding back on the seat of a Harley-Davidson with a man wearing a ponytail hanging down his back, and say, 'Dad, meet my husband,' I'll put out my hand and say, 'Welcome, son-in-law.'" I'll hate it, but I will do it because Jaclynn is my daughter. Prudent people stay with their commitments, their families, their spouses, their church, and their school.

Simple people are easily led astray. Simple people say, "You made one mistake, so I'm leaving the church," or "I don't like the way you treat me. I'm leaving home."

Both the simple and the foolish one lack understanding. Proverbs 8:5 says, *"O ye simple, understand wisdom: and, ye fools, be ye of an understanding heart."* Understanding is the mental grasp of a situation. It is the ability to distinguish or separate mentally that which is right and wrong.

I Samuel 16:17 and 18, says, *"And Saul said unto his servants, Provide me now a man that can play well, and bring him to me. Then answered one of the servants, and said, Behold, I have seen a son of Jesse the Beth-lehemite, that is cunning in play-ing, and a mighty valiant man, and a man of war, and prudent in matters, and a comely person, and the Lord is with him."* At this time, David was 17 years old, and he was called a man—not a young man. The term "young man" has nothing to do with age; *young man* is a characteristic.

The Bible says that 17-year-old David was *"pru-dent in matters."* The Hebrew word *prudent* is exactly the same word for *understanding* in Proverbs 8:5. A prudent man is someone who understands the big picture. David had understanding because he was a already a prudent man. The most common usage for *prudent* is understanding. *Prudent* describes someone who understands the big picture. He is able to see beyond the issues at hand and see the real issues down the road.

3. Unsuspicious. A simple person does not nat-urally suspect guilt or wrong. Simple people do not see the wrong or evil in people. Because of this inability, they are vulnerable to wrongdoers and shady people, especially the intelligent ones. They are vulnerable to people who are smart enough to know how to cleverly maneuver them. The prudent guide is the one who knows how to cleverly maneu-ver his young person to wisdom. The scorner can also be very clever, and he can maneuver the simple person because he preys on his gullibility and unsus-picious nature. An unsuspicious person does not nat-urally suspect guilt or wrong; he is prone to make

judgments without the supporting evidence.

The scorner preys on the unsuspicious nature of the simple one. He also preys on the fact that the simple one does not need any evidence of what he wants him to do. After all, the simple one just does not know. When confronted with wrongdoing and asked why, the simple one says, "I don't know." When asked, "What do you want to do with your life?" the simple one typically answers, "I don't know." Ask a teenager any question, "How was school?" "How was revival?" or "How was your bus route?" He answers with the three famous words of all teenagers, "I don't know." The truth is, he really has no idea. It requires a clever person to have an intelligent conversation with a teenager!

I systematically checked every definition of *scorner* and *prudent* in the Greek and the Hebrew. A prudent guide uses vivid, forceful, and persuasive words. He is eloquent. He recognizes and understands the bigger picture. He is very discreet and careful in how he answers and what he says when he answers. The word *prudent* literally means "smooth, clever, and artful." The word *deceitful* is never used to describe him, but he is extremely skillful in persuading people to get them where he wants them to go.

Prudent also means "intelligent," "an expert," and "to be able to teach and communicate." The prudent person is successful in his field; he is mentally well-wrapped.

As I already mentioned, he knows how to use words and is eloquent. Most think of eloquence as someone who uses $64 words, but eloquence does not mean that at all. Rather, an eloquent person is someone who paints pictures with vivid, persuasive speech. "Crafty in his counsel" is the literal meaning for *eloquence*.

The word *scorner* literally means to "make mouths at." It is easy to spot scorners during preaching as they make faces or roll their eyes. The scorner is a boaster and a braggart. He loves to talk about how good he is or how good he can do something. The scorner laughs in mockery to avoid humiliation when he is caught doing wrong. The scorner despises correction and receives it with contempt. The scorner mocks authority.

Unfortunately, parents are sometimes scorners rather than prudent guides, when they are trying to get their children to wisdom. They make fun of their children. They make faces at their children. They roll their eyes, and shoot off their mouths. I have watched in total disbelief as a parent berates, humiliates, scolds harshly, uses abusive language, mocks, makes fun of, ridicules, and holds in contempt the very simple person he is trying to take to wisdom. Years later, that same parent says, "Whatever happened to my child. How did I get a fool for a kid?"

Quite simply, that parent acted like a scorner instead of a prudent person when his position was supposed to be that of a prudent man. Then, your simple child came under the spell of a very prudent scorner—a slick, conniving, crafty individual who kept his mouth shut. I have watched scorners slowly, craftily, and artfully bring the simple to foolishness because they used better methods than the prudent parents. It is easy for the child of a scornful parent to get under the spell of a slick, conniving, slippery fool.

The Bible has provided us with the example of a scorner. His name was Jonadab. This friend of Amnon, David's son, was *"a very subtil man,"* who led Amnon to foolishness.

The word *subtil* in II Samuel 13:3 used to describe Jonadab, is exactly the same word used to describe the serpent in the Garden of Eden. Genesis 3:1 says, *"Now the serpent was more subtil...."* In the confrontation between God and the serpent and Adam and Eve, the serpent won. Adam and Eve were simple ones who had been untaught and untrained. They were simple blank pages. Adam was supposed to keep Eve out of trouble and lead her to wisdom. Instead, he went right with her to the slaughter. By his craftiness, the serpent persuaded Eve to sin.

A subtil man doesn't separate a young person from his parents by saying, "You're dad is a fool!" Rather, he subtly says, "You have a mighty fine dad, but he's a part of the older generation. He's a very smart man and a very successful man in his generation. Still, members of that older generation do not always see things our way. Surely your dad taught you to have your own opinion on certain matters. What do you think about...?"

We sometimes stand aside as prudent guides and say to our young people, "You punk! What's the matter with you? You're an idiot! Just be that way! Get away from me! I'm over here where wisdom is."

The prudent parent needs to use wisdom in dealing with his child and say, "Hey, Son, let's go for a cup of coffee or a Coke. You know, buddy, I'm with the older generation, and I know there will be voices who will tell you to be your own man and to think your own thoughts, but you're pretty young. I am going to share what my dad taught me about how to get wisdom. I'm going to be here to make this journey with you."

The truth is, if prudent parents were as crafty at right as the Devil is at wrong, we'd have more children turning to wisdom rather than are turning to foolishness. It is not that the scorners are so good at pulling away our children. It's that those of us who think we are wise are driving them from us. The smartest guides and the wisest leaders craftily pull and maneuver their followers to wisdom. The Bible says not to be ignorant of the Devil's devices; he is subtle. The Devil actually took the word that describes the prudent person—*subtil*—and used it to expertly maneuver Adam and Eve into being foolish.

Young people need to have their hearts hooked by a prudent guide. Their heart needs to become fastened to that person who represents wisdom. Then the simple one will buy into the prudent way of thinking. One way to capture the heart is to have fun with the simple ones. Go out for a bite to eat. Spend a little money on the simple one.

My goal is to get a whole generation to buy into the old-fashioned, fundamental way of thinking. Young people do not have to wait until they are 30 or 40 years of age to get wisdom. They can get it right away. Jesse must have been some father because David was 17 when he bought in. Jesse reared one prudent son. By the time David was 17, he was a mighty man of valor, a prudent man in all matters. That is exactly what I want the teenagers of First Baptist Church of Hammond to be.

CHAPTER V

The Lip and the Rod

"In the lips of him that hath understanding wisdom is found: but a rod is for the back of him that is void of understanding." (Proverbs 10:13)

*U*nderneath the whole concept of helping the simple person become a wise man lies one of the most important ingredients in getting him to choose wisdom. The prudent guide must capture the heart of the simple one. If the simple one does not have a heart for right or a heart for God, he will not choose wisdom. That is why Solomon said, *"Son, give me thine heart."* Though the choice for wisdom is not a purely logical decision, logic is involved. Though not a purely mental decision, some mental understanding is needed. There must be understanding and intelligence and logic involved when seeking wisdom, and there must be knowledge and information and facts used.

Unfortunately, the problem of seeking wisdom is never a logic problem; the problem is a heart problem. If the simple one would logic, he would find that common sense says, "I should go toward wisdom. That's where the happy people live. Wise people are those who have money; they are not in debt. They have happy marriages; they are married to the same person with whom they fell in love when they were young. Their children rise up and call them blessed. If everything logical tells the simple one where he should be, then why do so many people choose foolishness? The reason is because it's not a logical decision; it's a heart decision. Many logical people push a simple one over to foolishness by not capturing his heart.

The most delicate step in capturing the heart of the simple one is in the area of discipline. How you discipline or chasten a child is critically important in capturing his heart. Those who learn the art of disciplining and chastening and correcting a child have mastered one of the greatest heart-keepers of any technique they can have as prudent guides. In other words, learning to discipline the simple person is the greatest art in getting him to wisdom. At this singular time in a person's life, if this needed correction or even chastening is handled correctly, it can be one of the most important times in a child's heart development for God. If those times are not handled properly, the simple person is driven to foolishness.

Proverbs 10:13 says, *"In the lips of him that hath understanding wisdom is found: but a rod is for the back of him that is void of understanding."* The words, *"void of understanding,"* describe a person who does not have a heart for right—the simple person. Since he does not have a heart for right, he needs along with instruction and understanding a rod of correction.

If the prudent man has the simple one's heart, he will be able to lead him simply by what he tells him. Once the prudent guide has the simple one's heart, the simple one will do what his prudent guide tells him to do. Until the prudent guide has his heart, he will sometimes have to use stronger pressure and more effective measures to get his attention. The Bible says in Proverbs 17:10, *"A reproof entereth more into a wise man than an hundred stripes into a fool."* If the simple one doesn't listen and the prudent guide doesn't exercise enough wisdom to get him, the simple one will stray further from wisdom. By the time the simple one gets to foolishness, he can be spanked and spanked and spanked, but he will not listen. However, if the prudent guide, who is on the journey with the simple one, has captured the heart of his simple one, all he might have to say is, "Son, you know better." The heart of the simple one is crushed by the correction.

Proverbs 13:24 says, *"He that spareth his rod hateth his son: but he that loveth him chasteneth him betimes."* Please do not think that I am advocating hurting a child in any way. I do not want anyone to hurt any child. His feelings may need to be hurt, and God has provided a place on a child's anatomy for him to be properly disciplined where no bodily damage will be done. A spanking should take place at the proper time, in the proper place, and in the proper manner. The child's nerve endings may hurt, but it won't do any bodily damage beyond that. God is simply saying, "If you're not willing to go to the point where you absolutely have to inflict a certain amount of pain in a child, don't tell Me you love your child." A modern-day philosophy is permeating our society that says children should not be spanked. With that thinking, one might as well tell his child that he hates him. Keep in mind that every human being is a sacred person created in the image of God. Even those whom the government executes are given certain rights and dignities. With that thought in mind, be very careful in how you discipline your children.

Society would be in chaos if we said, "We're going to do away with policemen and not give any more tickets. Don't speed. We won't throw you in jail if you get in an accident. We'll just slap your hand." That won't work; society must have enforcement. That is why policemen have weapons. In society, we must have laws which have to be enforced. If one is broken, punishment has to be meted out.

In the area of discipline, there must be a certain boundary that cannot be crossed. If a child chooses to cross that point, he will be in trouble. The mom and dad who say, "Well, I don't believe in spanking my children," have taken away a great deterrent for doing wrong. Proverbs 1:7 says, *"The fear of the Lord is the beginning of knowledge...."* That fear factor is a way to capture a simple person's heart at a young age. Spanking, not child abuse, is a biblical concept.

Proverbs 22:8 says, *"He that soweth iniquity shall reap vanity: and the rod of his anger shall fail."* I believe this is a key verse regarding why people spank poorly or do not spank at all. Chastising or punishing in anger is often evidence of an authority who is in error himself. The most violent of people are those who are guilty, and they fear that the sin of which they are guilty will be discovered. Their anger is a clear sign of their not having dealt with that sin.

When a parent lashes out in anger and beats a child or hurts a child, he is admitting that he himself is out of control. That parent is guilty of a sin which is using his child's body to exercise "self-inflicted" punishment. That kind of punishment is very wrong.

Perhaps a prudent guide has done something wrong in the past. Suppose he sees a simple one going in the wrong direction. If he is not careful, he will abuse the simple one as a way to punish himself for what he should have been punished for when he was younger. The Bible says that this man sows iniquity; and when he spanks the simple one, it will bring no lasting fruit. The effect will be vain and empty, and the rod that he uses against the simple one will fail. The simple one will not be brought to wisdom; rather, he will be driven to where the prudent person does not want him to go—foolishness.

I have counseled with many parents who tell me they spank their children, but the children still do not do right. I ask them: "Do you spank with a violent temper?" or "Do you spank in anger?" Parents who are in control of themselves never hit a child in anger. Don't ever deal with any situation when your anger is out of control.

If a child needs spanking, never grab him in anger. When a child is grabbed, pushed up against a wall, or the parent yells, "If you do that again, I'll tell you what...." you are reacting with a violence that says, "I'm out of control." God says the anger and chastisement brought to the simple one at that time is vain. The rod will fail. It will not accomplish what you want.

Using good methods when spanking will always knit the heart of the simple one to the prudent guide. Proper spanking always results in the one being disciplined looking at the authority and saying, "Thank you. You're the best friend I've ever had." That is the way it has to be. Spanking does not bring the desired results; it is your heart understanding with what that child needs. If you do not understand where you are going, the purpose for which you are spanking, and where you are bringing that child, it's not time to spank that child.

Unfortunately, people who hurt children by threatenings or beatings give good reason for the government to get involved. People who are drunk and on drugs have no business spanking children. No one has the right to beat and abuse a child. That kind of correction is not scriptural spanking. Because of the way a simple one is disciplined, it is possible to spank and drive him in the direction you did not want him to go. A wise parent must consider, "Am I angry at myself because my child is doing what I did as a child? And because I did not get caught doing it, I'm never going to let him do that?" Those are wrong reasons to spank.

Some parents who turn out the best kids had a pretty rough childhood. Certainly, no one set of parental standards turns out good kids. Some people who have wicked, vile pasts have turned out wonderful children. On the other hand, some people who have spotless paths that are clean as stainless steel, have children who have broken their hearts. No magical standard rears good kids.

Some people who have never been spanked are wonderful, model Christians. Some people who have been spanked are miserable people. The key is not the spanking or lack of it. You have to be willing to spank your child or to use force, if necessary. But if force is used when the heart is angry and hot, that force will have been used to drive the child in the opposite direction intended.

Proverbs 22:15 says, *"Foolishness is bound in the heart of a child; but the rod of correction shall drive it far from him."* The word *foolishness* is the same as "silliness." A simple person is silly. The Bible says we must be able and be willing to use physical punishment if necessary to drive that foolishness or silliness out of the heart of the simple one.

Proverbs 23:13 says, *"Withhold not correction from the child: for if thou beatest him with the rod, he shall not die."* This verse cannot be taken out of context. It is not talking about hitting and hurting a child. The word *beatest* means "to hit him with instruction." In a sense, the Bible is saying, "Son, I know what you need. I know you need a spanking now, and I'm willing to apply the board of education to the seat of learning to the degree necessary to make the head understand that the bottom speaks." When a child is properly spanked, somehow the result makes their brain cells begin to think very well! God is not talking about hitting until a kid is hurting. Be very careful how you interpret these verses.

I do not advocate in any way the beating of children. Beating a child is wrong! I do advocate corporal punishment if it is administered correctly. A pastor said to me recently, "How do you get by with corporal punishment in your schools?"

"Because we have a system," I said. "We have a procedure that keeps us from getting in trouble. That is the whole secret."

Proverbs 26:3 says, *"A whip for the horse, a bridle for the ass, and a rod for the fool's back."* The rod is usually designated for the person who is well on his way toward foolishness. A rod is never mentioned with a wise person. Once a child's heart has been reached, the spankings should greatly diminish. If a child is being spanked many times, his heart has not been reached. He is not being steered toward wisdom. The further a child progresses toward wisdom, the fewer spankings he will need. The Bible has no verses that deal with beating. The rod is for a person who acts like a fool. The rod is for the scorner. In fact, the Bible says in Proverbs 19:25, *"Smite a scorner, and the simple will beware...."*

The Bible says that a horse needs a bridle to make him respond properly. In much the same way, a fool needs a rod to make him respond properly. When a simple one is headed toward foolishness, the prudent guide sometimes needs a rod to draw him back in the right direction. When a simple one is headed toward wisdom, the lips or the words of instruction lead that simple one. The prudent guide can just look at the simple one, and he will do right. As a general rule of thumb, the more you are spanking your child, the more you are admitting that your child is bending toward foolishness and that you do not have his heart. The more you have his heart, the less you have to apply the hand. It is very important to understand this concept.

I am not saying that the simple one, who is on the path to wisdom, doesn't need some spanking. Staff members need some correction once in a while. Adults need some firm correction sometimes. At times, the pay of someone who is misbehaving has to be docked. The hand or rod should be used against them very rarely.

When one of my staff members made a mistake, there was a memo underneath my office door explaining the mistake when I got back to my office. This person apologized profusely. I wrote on the bottom of that memo, "When staff members like you admit their mistakes and punish themselves, I don't have to." No correction was necessary. A truly wise person punishes himself for his mistakes, and the prudent person doesn't have to punish him. The fool is not smart enough to recognize his own mistakes.

Correction of simple ones who are on their way to wisdom gets less corporal and needs less of the hand because the lip is effective. Correction of the foolish is more of the hand and more of the rod, and the lip is ineffective.

Proverbs 29:15 says, *"The rod and reproof give wisdom: but a child left to himself bringeth his mother to shame."* Both the rod and reproof are needed to produce a wise child. The *rod* is "corporal punishment," and *reproof* is "the words of instruction." Physical correction supported by instruction is the right combination, and a prudent guide needs both of them.

If my only form of child rearing is a swat on the bottom, that is using only half of the formula, and it will not turn out the child I want. It may slow the deceleration of life or the decline of morals, but it will not get him to wisdom. On the other hand, if all I do is lecture, that will not work either. Again, only half the formula is being used. The Bible says the rod **and** reproof give wisdom. You must have both instruction and correction; both are necessary.

CHAPTER VI

Preparing for the Race of Life

"Wherefore seeing we also are compassed about with so great a cloud of witnesses, let us lay aside every weight, and the sin which doth so easily beset us, and let us run with patience the race that is set before us. Looking unto Jesus the author and finisher of our faith; who for the joy that was set before him endured the cross, despising the shame, and is set down at the right hand of the throne of God." (Hebrews 12:1, 2)

*N*othing drives the simple ones away from the wise, prudent guides more than the misunderstanding and mishandling of discipline. By discipline, I do not mean just a spanking; I am referring to the whole business of the correction of an error, the instruction of how to do things right, and the constant admonishing to do right.

The concept of chastisement is a little more complicated than just grabbing a board and spanking a child. Hebrews 12 is the chastisement chapter of the Bible, although there are other references to this subject in the Bible. The word *chasten* only occurs six times in the whole Bible, but the words *chasteneth* or *chastened* occur many times. In Hebrews 12, the chastening and discipline chapter, the first verse begins with the word *wherefore*. To understand the concepts in Hebrews 12, we must start in Hebrews 11.

"Now faith is the substance of things hoped for, the evidence of things not seen. For by it the elders obtained a good report. Through faith we understand that the worlds were framed by the word of God, so that things which are seen were not made of things which do appear. By faith Abel offered unto God a more excellent sacrifice...." (Hebrews 11:1-4)

It is very interesting to note that these verses did not start with Adam or Eve or Cain; rather, the first person mentioned in Hebrews 11 is Abel. Many important people are mentioned: Abel (verse 4), Enoch (verse 5), Noah (verse 7), Abraham (verse 8), Sara (verse 11), Isaac (verse 20), Jacob (verse 21), Joseph (verse 22), and Moses (verse 23). In verse 30 the walls of Jericho are mentioned, and this reference points to Joshua. Rahab the harlot is mentioned in verse 31. Gedeon, Barak, Samson, Jephthae, David, Samuel, and the prophets are mentioned in verse 32. Then the chapter describes people like Daniel without mentioning them by name. Hebrews

11:39, 40, and 12:1 say, *"And these all, having obtained a good report through faith, received not the promise: God having provided some better thing for us, that they without us should not be made perfect. Wherefore seeing we also are compassed about with so great a cloud of witnesses, let us lay aside every weight...."*

Hebrews 11 lists the names of many whom God calls witnesses in Hebrews 12:1, *"Wherefore, seeing we also are compassed about with so great a cloud of witnesses...."* The word *witness* is a Greek word that can mean "somebody who testifies" or "somebody who dies on behalf of." It is the Greek word *martyr*. So a martyr and a witness are the same. They give testimony to something, or their life testifies to something. The way they lived their life gave testimony that God said, "I want you to mark that testimony." Some of these witnesses even went so far as dying to make a statement by their lives. These people mentioned in Hebrews 11 made a statement by their life. God was saying, "Mark Abel, mark Noah, mark Abraham, mark Sara, mark Isaac, mark Moses, mark Barak, mark Gideon, and mark Jephthae. God said mark them because the lives of those people bear testimony. Their lives have testified or witnessed all the way to death, and in some cases, they have testified about something very important.

Hebrews 12:1, 2 tells us these are people who ran and finished the race God had set for them. Many people whose lives were recorded in the Bible ran a good race, but they did not run their race to the finish line. Many of the people in Hebrews 11 ran a race and even got off the track for a while, but the people who were mentioned got back on track and finished their race.

Many famous people are not mentioned in Hebrews 11. I think King Saul is not mentioned because he started the race, got out of the race, and never got back on the race track. A lot of good people ran the race for a long time, but they veered off the course, stepped outside the line, sat down, were pulled out, or for some unknown reason, they stopped running their race, and they never got back into their race. Some people, like Samson, ran the race, stepped out of the race, got back in the race, and finished the race stronger than when they started the race. When Samson left the race, God chastened him and brought him back in the race. I believe Samson finished his race stronger than when he started, that is why God placed him in the "Christian Hall of Fame."

People often ask me why those particular people are mentioned in Hebrews 11 and not others. No doubt there are other people in the Bible who could have been mentioned, but God wanted to highlight a select few. I believe those mentioned are the classic examples—the "Hall of Famers." They are the long-term people who entered the race that God had chosen for them and stayed in the race all the way to the end. When they stepped back in after being chastened, they strongly finished the race that God had set for them to do. Now that they are the spectators watching us, let us use their lives as an example of how to run the race and how to stay in the race. If perchance we get out of the race, let us respond as they did when God gently or forcibly pulls us back in the race.

Abel was the first human being not to swerve aside. Adam and Eve got out of the race and never got back in that race. Cain started the race and never finished the race. Abel was the first human being who started the race, stayed in the race, and finished. As the first martyr, his life was cut short because of his testimony. His witness bears evidence that he believed in the sacrifice that God would send. He was the very first person whom God counted all the way to the end. Abel finished the race that God gave him to do.

We have a *cloud* of witnesses in Heaven. A cloud in the Bible speaks of "a crowd." When people have died and gone to Heaven and they're all together, they are called a cloud. On earth, they are called a crowd. They are spectators who bore witness through their lives, whether short or long, that they started a race and finished.

Hebrews 11 is telling us about this group of people. They began a race that God specifically had for them. Hebrews 11:13 adds, *"These all died in faith, not having received the promises, but having seen them afar off, and were persuaded of them, and embraced them, and confessed that they were strangers and pilgrims on the earth."* According to this verse, these people who were in a race saw "a land that is fairer than day, and by faith they could see it afar." By faith they saw a goal.

They said, "At the end of life, there is a blessing

for me. At the end of life, there is the promise that God will abundantly reward those who diligently seek Him." So, all of these people got in the race and said, "I believe if I get in my race, stay in my race, or at least finish my race, God will greatly reward me." These are the people who believed God's promise, and by believing that promise, they ran their race. The word *run* means either "to walk hastily" or "to move quickly along." These giants of the faith finished their race, believing that God was going to reward them. This introduction is the background for the chastening chapter—Hebrews 12.

In Hebrews 12:1, we read about those who have gone to be as spectators or examples. Some were martyred, but all witnessed through their lives the race of God and God's blessings and reward upon them.

Hebrews 12:1 also says that we run a race, *"...let us run with patience the race that is set before us."* God is saying, "Let us as a team run the race." So, we run with them.

Hebrews 11:39 says, *"And these all, having obtained a good report through faith, received not the promise."* God said, "I've set up a long race that goes all the way back to Abel and continues all the way to this generation."

He says to every believer, "You and I have a race to run." "You and I" has a double meaning. First of all, the race that each runs is a continuation of the race that the cloud of witnesses began running way back with Abel, and we are to run the race with them. So when the race is run, each had better run that race the same way they ran it. In other words, they ran for the will of God. They ran for the glory of Jesus Christ. They ran for the blessings of God. They ran and did not turn to the left or the right. Just like they did, we had better find our place in the will of God.

What is the will of God? Start with Abel and study everyone listed in Hebrews 11. You will know the will of God for your life. It's right in line with those people. It's a life of faith, trusting God, believing God, believing His Word, and obeying His Word, just as those people did. Really, the will of God is to obey the Word of God. God is more interested in the **what** of His will than the **who** and the **where**. We get all bent out of shape in the **who** and the **where** of God's will and forget about the **what** of God's will. When a young man is in the will of God, he is in the same lineage as Barak, Gideon, Jephthah, Moses, Abraham, Noah, Enoch, and Abel. That's exactly the lineage in which every simple one should be.

This race of life is a relay race. Abel ran a little while, and when his race was cut short, he handed the baton to Enoch who grabbed it. Enoch ran for a while and handed it to Noah. Noah handed it to Abraham. Abraham handed it to Moses. Moses handed it to Barak, who gave it to Gideon, who gave it to Jephthah, and on down the line. That baton continues to be handed down. Everyone has a will of God for his life with the same baton still being used and the same race still needing to be run. That is the will of God for a Christian's life.

"Let us run" means that we want to run the same

race that these "Hall of Famers" ran. When a Little League baseball player picks up the bat, he wants to be like his hero who is in the Hall of Fame. His hero might be Reggie Jackson, Mickey Mantle, or Barry Bonds. He will choose a hero, and he's going to say, "I'm going to swing the bat like Sammy Sosa of the Cubs!"

The race that started back with Abel is still going on, and we are now in the race. That means that each one of us should have for our heroes those people listed in God's "Hall of Faith." There is a way to make Barak and Gideon our heroes. When they were tempted, their besetting sin tried to hold them down, but they did not let the weight hold them back. Use these heroes in God's "Hall of Faith" as examples for when your sin besets you. The word *beset* means "when it surrounds you." I Corinthians 10:13 says, *"There hath no temptation taken you but such as is common to man: but God is faithful, who will not suffer you to be tempted above that ye are able; but will with the temptation also make a way to escape, that ye may be able to bear it."*

When temptation to quit the race faces you, say, "I'm tempted to get out of my race, but So-and-so did not, and So-and-so did not, and So-and-so did not, so I guess I won't either." That is what this verse is saying. I Corinthians 10:13 is also an example of when someone does step out of line, God will do something to get him back. God will spank him. The reaction of the simple one will be, "I don't appreciate that," and he will be tempted to go further away. But if he will use men like Samson as examples, the simple one will get back in line. When God spanked Samson, he got back in line. If God spanks you, then get back in line. Use those biblical examples of how to run the race; use them as your mentors or heroes.

If a Little Leaguer goes through a batting slump, his dad might say, "That's all right! Even Frank Thomas goes through a batting slump."

"I know, but Frank Thomas is a hero," the simple one might say.

The prudent dad says, "Where do you think Frank Thomas started? He started when his daddy bought him a baseball bat when he was a little boy, and he started swinging."

"He did?" asks the simple one.

"Yeah!" exclaims the prudent dad.

"Wow, you mean he really goes through hitting slumps?"

"He sure does!"

"Okay," the simple one says, "then I'm going to make him my hero. I'm going to get back in that batter's box and swing harder."

In the same way, the people in Hebrews 11 can be used to inspire us to fulfill the race that we have before us. We should study their lives.

Proverbs 19:18 says, *"Chasten thy son while there is hope, and let not thy soul spare for his crying."* This is a beautiful verse. On the surface, it seems to say, "Go ahead and spank your child while you still have hope of redeeming him, and if he's weeping and wailing, hit him anyway." I do not necessarily believe that is a bad interpretation, but I do not think it's an accurate interpretation.

The Bible teaches that the word *hope* means "while there is a cord attached between the two of you" or "while the two of you are still connected." It literally means while your hearts are still connected. Chasten your son when it is going to do some good.

The verse continues, *"...and let not thy soul spare for his crying."* The word *crying* means "to die." The whole verse means, "If I will chasten my son while chastening does any good, while he and I are still heart-to-heart with each other, then I will have no grief if I lose him."

If he dies prematurely, it will not be a grief; it will be a sadness. The greatest grief is a child whom God has to spank and spank and spank because the parent did not. That child never gets back in the race. When he never gets back in the race, your soul is carried away in the loss of your son. If a parent will discipline his child while it will do some good or while the cord is still connected between the two of them, then the parent will have no regrets because his child will be back in the race.

He is saying, "Son, we are going to walk or run our race. The race is yours and mine. It is a father and a son walking the race together. While we are walking the race together, and my heart is knit with your heart, and your heart is knit with my heart. If you're tempted to turn out of the way, then I can pull you back in because my heart and your heart are still connected. It will do some good." He's saying chastening really only does any good when there's already an invisible "heart cord" attaching those two hearts.

If you lose the heart of your child, you can beat him until there's nothing left of him, but it will do no good. God says, "Your soul will be taken away with the loss or thought of his producing an unsuccessful, unproductive, out-of-the-race kind of a life." What a heartache to watch a child walk out of the race! God is saying that the pain of spanking your child back into the race is nothing compared to your soul's feeling if he gets out of the race completely. Your soul will be carried away in the overwhelming grief, and you will go through your race almost feeling that life is not worth living. "My soul will have died," is what it is saying.

A son who breaks that bond and a son who cannot be brought back causes his father's soul to feel like David when he said about his son Absalom, "...*would God I had died for thee, O Absalom, my son, my son!*" (II Samuel 18:33) Was David dead? No, he said, "My soul is so overwhelmed with grief that I'd rather die than run this race by myself. I'm running my race because I believe there are promises for me, but those promises really don't matter much if I can't share them with the people I love. I really want the promise to be shared, but if my heart is not with my son, my words sound like meaningless noise."

We have been taught about the heroes of the faith like Abel, Enoch, Noah, Abraham, Moses, Barak, Gideon, Samson, Jephthah, Daniel, David, Joshua, Brother Hyles, J. Frank Norris, Billy Sunday, and John R. Rice. Dr. Lee Roberson is still running his race. We are still running the same race they ran.

One thing the Catholics have on fundamental

Baptists is that they have captured the market on heritage. They will say, "Grandma was a Catholic, grandpa was a Catholic, mom was a Catholic, and dad was a Catholic. I was Catholic-born, Catholic-bred, and when I die, I'll be Catholic-dead." I often wish we would capture that kind of loyalty in our Baptist churches. I wish we would look back and realize that our heritage does not go back to a pope; our heritage goes back to heroes like Abel, Noah, and Enoch. Our Baptist heritage goes back a lot farther than the Catholic church that was started in 313 A.D. We go back to 4,000 B.C.—6,000 years ago. We go back to when God said, "Let there be...." A race has been run that goes all the way back to 4,000 B.C., and we are still running that race. "Let us run the race" means that those of us who have been in the same lineage run it with the same people who are followers and leaders right now.

This whole chapter on chastening in Hebrews 12 is built on the idea of a father and a son, or a mother and a daughter, or a parent and a child, or a coach and his boys, or a Sunday school teacher and his children, or a leader and a follower running that race. It's particularly the parent and child running the race. It is my job as the leader to make sure you stay in the race. If you step out of it while our hearts are close, I can help you get back in line before I lose your heart. God lets me chasten or discipline you to bring you back in line.

Proverbs 19:18 says, *"Chasten thy son while there is hope, and let not thy soul spare for his crying."* The chastening is only going to be good while my heart is tied to the heart of the son that I'm trying to instruct in the ways of the Lord.

God said that while you're running this race you have two great enemies. The two great enemies are found in Hebrews 12:1, *"...let us lay aside every weight, and the sin which doth so easily beset us...."* The weight is the baggage that we call the guilt and remorse of having turned aside from our race. You turn aside, get out of the race, and then you go and sin. When you get back in the race, you have the baggage of, "I don't deserve to be in the race," or "God can never use me again," or "I no longer deserve this," or "You don't know what I did." You will carry the baggage of the guilt and the remorse of swerving aside.

God says, "Don't sin because when you do, it's going to bring baggage. If you do get back in the race, the baggage will be very hard to shed."

Many people come to my office and say, "Brother Schaap, I don't deserve to be used of God." No one has ever deserved to be used of God! It's always been by grace. A Christian entered this race by faith, and those heroes in Hebrews chapter 11 got in the race by faith. If they got out of the race, they got back in it by faith. They ran the race by faith, they lived it by faith, and they stayed there by faith. They got right with God by faith, and they got protected by faith. They were kept by faith.

Baggage or weights are those unnecessary emotional bags you pick up. It is a simple one's saying, "I hardly can walk through life. You won't believe what I did."

It's Samson's saying, "There's a Delilah in my life."

It's David's saying, "There's a Bathsheba and a Uriah in my life."

It's Moses' saying, "But I murdered a man, and I smote the rock twice."

It is people whining and griping about life. Weights keep people out of the will of God as much as the sin does.

Again and again people say, "You don't know the bags that I carry with me. You don't know what my past was like." Your past was Hell! You came into this world heading for eternal damnation. You didn't get any better. Your flesh hasn't improved a bit. Nobody's gone from 0 to 80 looking better all the way; rather, he peaks about 45.

In counseling, again and again I see a young lady who's godly and pure until she meets a boy. Maybe they misbehave, or he's not as good a guy as she should have, but she loves him. Perhaps she does something wrong or disobeys or gets in a little trouble. She says, "Well, it's probably what I deserve. I don't deserve the kind of good boy my mom and dad want me to have. I might as well settle for what I deserve!"

I beg every young lady, "Don't you do that! Don't settle for what you think you deserve. Settle for nothing less than the perfect will of God for your life."

You say, "But you don't know where I came from!" I know where you came from! You came from your mother's womb headed for Hell. You got here by faith, and so you've got to live by faith.

The second enemy is "sins" that easily beset me.

That means certain sins tend to come around and surround me. Those sins are attracted to me, and I pull them like a magnet attracts metal filings. God says, "Cast those besetting sins aside and push them aside." He wants us to push them aside and keep going. You say, "I can't do it!" Yes, you can because the Bible says in I Corinthians 10:13, *"There hath no temptation taken you but such as is common to man...."* God says that He will never let a sin get a hold of you that is more powerful than your ability to say, "No."

The question is asked, "How do I push away my besetting sin?"

I can teach everyone a marvelous, scriptural way to get rid of sin. You say, "NO!" Many of us, when we have temptation come our way, we do not say, "No." Instead, we say, "Well, let me check it out."

Instead of checking it out, say, "NO! NO! NO!" God will part the sin like the Red Sea, and you will walk through.

Go ahead and say, "I don't believe that." Try it! God will not show you the way through until you say, "No," to the sin.

When Moses got to the Red Sea, and God said, "Go forward," Moses said, "I can't because of the Red Sea." God said, "I said, 'Go forward.'"

Moses had to actually march forward before the waters began to part. When the children of Israel came to the Jordan River, God said that the priests were to put the Ark of the Covenant on their backs and walk into the river. The river would not part until they walked forward. Just like the children of Israel stepped forward by faith, each of us also has to take a step. While we take the step, we may be saying, "No, I don't want to do this." However, when we do step forward, God parts it. Then we realize we have the grace of God to go through our temptations without sinning.

God says there are two big enemies against me in running the race. The first is the sin that so easily besets me. If I submit to that sin, the second big problem I have is the weight that I will have on my shoulders.

CHAPTER VII

The Philosophy of Chastisement

"And ye have forgotten the exhortation which speaketh unto you as unto children, My son, despise not thou the chastening of the Lord, nor faint when thou art rebuked of him: For whom the Lord loveth he chasteneth, and scourgeth every son whom he receiveth." (Hebrews 12:5, 6)

*A*s a father and son walk through life, the son faces many temptations. One temptation is to go in a different direction. At that time the dad has to gently pull him back in. If you don't keep that close bond and let your son get too far away from you, he will wander and wander; at some point, he will be beyond your reach. His heart is too far from you. You can yell and scream, throw a fit, and you can say, "I'm going to kick you out of my house." But he is going to say, "It doesn't matter." He has gone too far. At some point, he is going to feel he has gone too far and then judge himself as unworthy of being back close to his dad's side.

The reason why I don't want people to go into sin is that there is a point they will get to where they will feel that they are no longer welcome back to the intimate relationship they not only had with mom and dad, but with their church, their pastor, and even with God. You can never wander too far that God won't take you back. Never!

However, what is going to happen is the simple one will feel the repercussions. He will go beyond our help. When the heart separation is that great, he is too far gone. We want to keep things close. The moment the simple one turns away, grab him. Keep that heart real close. Chastening is that turning away and that grabbing back. A father's chastening might be gentle, a little firmer, or real firm. The degree of firmness to use is an act of wisdom.

Hebrews 12:5, 6 say, *"And ye have forgotten the exhortation which speaketh unto you as unto children, My son, despise not thou the chastening of the Lord, nor faint when thou art rebuked of him: For whom the Lord loveth he chasteneth, and scourgeth every son whom he receiveth."* The words, *"my son,"* show affection. The words, *"whom the Lord loveth"* and *"every son whom he receiveth"* also express endearment. God uses intimate, affectionate terms. God is saying, "I have My son whom I love and whom I receive." The language used in these verses means He is pulling closer and closer to that person.

God is not just speaking to "a son"; He is speaking to, "My son." It's not just, "My son." It is, "My son whom I receive." It's not just, "My son whom I receive." It is, "My son whom I love to receive."

Hebrews 12:7, 8, *"If ye endure chastening, God dealeth with you as with sons; for what son is he whom the father chasteneth not? But if ye be without chastisement, whereof all are partakers, then are ye bastards, and not sons."* Again notice the intimate, affectionate language used in these verses. We have a father and a son who have a bond together.

God is saying, "You are not just a son. You are My son—My personal son. You're not just My son. You're My son whom I love. You are not just My son whom I love. You are My son with whom I love to be. I receive you. I love to be with you." That intimacy, that affectionate relationship is what I am addressing because the whole basis for chastisement is built on the intimacy. To the degree that there is a bond of affection between the prudent guide and the simple one is the degree of effectiveness the chastisement will achieve.

God works most on those who seek Him most. A coach invests more time in a player who responds to him more than a player who doesn't want to show up for practice or put forth any effort. At some point, a coach looks at his players and says, "I've got you guys on my team, but I'll be honest with you, if you guys don't want my attention or my time, I'll let you stay on the team, but don't expect much playing time. If you are going to play on the team, don't expect me to give you much time unless you want to put forth the effort."

One player says, "Hey, coach! What about this?" He's the player who puts forth the effort and the energy. The coach seems to focus on that person and says, "I think I can invest more time in him because my investment is going to pay off." A musician or a teacher or a mentor invests more time and effort in the student who loves him. He shows he is a wise investment by his response.

II Corinthians 6:17, 18 say, *"Wherefore come out from among them, and be ye separate, saith the Lord, and touch not the unclean thing; and I will receive you, And will be a Father unto you, and ye shall be my sons and daughters, saith the Lord Almighty."* God is saying, "I would like to be closer to you than I am. However, you run with the wrong crowd. Come out from among them and leave them alone." If the person steps out from the wrong crowd, God says, "Now I can receive you a little more. I can do more work on you." God is not saying that He hates these people. He is simply saying that while a person is in that crowd, whatever investment He makes in that person is going to be squandered. God is not going to get a good return on His investment. God is saying, "For the amount of time I spend on you, the wrong crowd is going to undo all the good that I put in you."

You can be the best dad in the world or the best mother in the world, but if you have a son or daughter who runs with the wrong crowd, that crowd will undo all the good you put in your son or daughter. It's not because you hate the wrong crowd. Sometimes we drive our child to the wrong crowd by making him feel sympathy for the wrong crowd. We bicker and fuss and yell and fight so much about the wrong crowd and say, "I hate your friends, and I hate the people you are running with!" We condemn the child, and then he feels that the only people who accept him are the guys in the wrong crowd who do make him welcome. That wrong crowd is more shrewd, brilliant, clever, and prudent than the wise guide as they make statements like, "We're not judging your old man. He's a good guy."

The wise guide needs to look at the wrong crowd and say, "I don't have a problem with them, but if you would leave them, I've got something that is better than what they can offer you. I've got me."

You need to understand the concept that God says, "If you leave that wrong crowd, I'm not just trying to get you to be a separated, fundamental Christian who walks around and condemns others." God is not trying to get you to turn around so you can point your finger at the wrong crowd and say, "Shame on you."

Sometimes the mentality we have is that we are to be separate so that we can look down our long noses at everyone who is not like us. That is not the purpose of separation. God says, "I have things to invest in you. I will teach you how to handle your money, but I don't want to do that if you are going to squander it on drugs, nicotine, alcohol, and foolish living. I've got wisdom to teach you how to have a happy marriage, but I'm not going to give it to you if you are running with harlots and with all the girls the wrong crowd wants to introduce to you. That is squandering what I want to give you. So, I'm not going to invest in you because if I do, it will be a waste of time. If you will leave that wrong crowd and show me some interest, then I will receive you. I will be a Father unto you, and you will be my sons and daughters."

God works most on those who seek Him most. James 4:8 says, *"Draw nigh to God, and he will draw nigh to you...."* God says, "I will match you step for step. You move toward Me, and I'll move toward you." God is not saying that He does not want to be close to you. He is saying that He cannot help you when you are with the wrong crowd. It is not that He doesn't like to be with you, but He can't rub off on you while you are in the wrong crowd. You are too distracted by a crowd that is keeping you from becoming what He has for you. He has more money than the wrong crowd. He has more wisdom than the wrong crowd. He has more wherewithal to build your life. When you're with them, He can't do that. So, if you want to be near God, take a step, take another step, and he will draw nigh to you.

John 1:11, 12 say, "He came unto his own, and his own received him not. But as many as received him, to them gave he power to become the sons of God, even to them that believe on his name." God said, "I came to My own, the Israelites, and they did not receive Me; so, I am so hungry to build a relationship, I'll open up a door to anyone—Jew or Gentile. Anyone who receives Me, I will receive him, and I will make something available to him. I'll make it possible for him to have a son-Father relationship with Me even though he is not My child. I chose the Israelites, but they didn't want Me. Not only did they not receive Me, they rejected Me. Not only did they reject Me, they crucified Me. They said, 'We will not have this man to be our king. We have no king but Caesar.' " What an incredible insult! So, God said, to all mankind, "Not just My Son, but to any of you who would like to be close to Me, draw nigh to Me, receive Me, and I will receive you. Who wants to do it?"

Jesus came to His own, and they rejected Him. Therefore, God said that He would give the authority and the privilege to be received by God and open it to all who believe. However, not all who believe draw nigh to God. Many people stay in the wrong crowd and say, "I believe, but I am not going that close. I'll receive you as my Saviour, but I don't want anything closer than that. I don't want you as my Father. Don't get too close to me. Let's just keep our distance. Let's be related, but let's not get intimate with each other."

Philippians 3:17-19 say, *"Brethren, be followers together of me, and mark them which walk so as ye have us for an ensample. (For many walk, of whom I have told you often, and now tell you even weeping, that they are the enemies of the cross of Christ: Whose end is destruction, whose God is their belly, and whose glory is in their shame, who mind earthly things.)"* God says that there are people who don't want to receive Him. There are people who say that they will receive God as a Saviour but not as a Father.

I might be a saved individual, but I might not have the intimacy that another Christian has. The Bible is full of examples of people who trusted Christ but got no further in their Christian walk than that. A lot of false doctrine is taught because some people assume that everyone who gets saved automatically wants to quit smoking, drinking, dancing, and fornicating. However, that is not true. You don't get rid of the old nature when you get saved! You get a ticket to Heaven! *"But to as many as received him* [as Saviour] *to them gave he power...."* If you want, you now have the privilege to get closer. If you want to be close to God, you can; but a lot of people don't.

Paul said, "I told you before, and now I tell you weeping, there are many that don't walk like that. They have even followed another god called the god of their human appetites. They have caused great shame and are the enemies of Christ."

II Thessalonians 3:6-8 say, *"Now we command you, brethren, in the name of our Lord Jesus Christ, that ye withdraw yourselves from every brother that walketh disorderly, and not after the tradition which he received of us. For yourselves know how ye ought to follow us: for we behaved not ourselves disorderly among you; Neither did we eat any man's bread for nought; but wrought with labour and travail night and day, that we might not be chargeable to any of you."* Paul teaches us that there are brothers and sisters in Christ who don't want to be close to God. They have received Christ to a degree, but they don't want to draw nigh to Him and have God draw nigh to them.

II Timothy 1:14 says, *"That good thing which was committed unto thee keep by the Holy Ghost which dwelleth in us. This thou knowest, that all they which are in Asia be turned away from me; of whom are Phygellus and Hermogenes."* Paul says that there are two men, Phygellus and Hermogenes, that are brothers in Christ. They were part of the whole group that embraced Paul, but they have pulled away and drifted back. Paul said they were no longer receiving him.

God said that I can be saved and related to Him, but that does not necessarily mean that I have an intimate relationship with God. What does an intimate relationship have to do with chastening? Hebrews 12:6 says, *"For whom the Lord loveth he chasteneth, and scourgeth every son whom he receiveth."* God says that He scolds every child. Verse 8 says, *"But if ye be without chastisement, whereof all are partakers, then are ye bastards, and not sons."* God says that He will chasten all the saved. God says, "All of you are going to get a scolding if you don't do right. But the ones who draw nigh to Me will feel the chastening rod more than the others." God will work more on the man who draws nigh to Him than on others, even though all may be the children of God. God chastens those He loves—the ones who draw closest to Him. The closer one gets to God, the more his conscience is smitten. The more God shames us when we do something wrong, and the more God is likely to say, "Listen to Me! I'm talking to you!" God always works most on those who choose to be closest to Him.

God chastens all of His children. The word *chasten* means "to instruct or to train or to punish." However, He chastens more those who seek Him the most earnestly. The simple one who draws nigh to the prudent guide receives the greatest investment. The simple one who walks away from his prudent guide may find himself quite alone.

Let me illustrate what I am talking about. Luke 15 gives the story of the prodigal son and his brother. The younger son says in verse 12, *"...Father, give me the portion of goods that falleth to me. And he divided unto them his living."* The younger son has the body of a man and the mind of an adult. The younger son says to his dad, "I'm splitting the scene. I'm out of here." This boy is saying, "I don't want to receive you; I just want your goods. Give me what you've got that is mine, and I'm leaving here." The dad did as his son requested.

Verse 13 says, *"And not many days after the younger son gathered all together, and took his journey into a far country, and there wasted his substance with riotous living."* Nearly everyone knows the rest of the story. The prodigal's father did not chase his boy. He let him go.

When the prodigal came back home, which boy was chastened and scolded by dad? Which boy received a tongue lashing from his daddy? He chastened the one who stayed home and didn't run off! Dad already knew it would do no good to chasten the younger son. "He doesn't listen to me anyway. I'm not going to waste my time trying to bring him into further nurturing." The prodigal's father said to the older son, "I thought I lost the boy. Now he is back home. I'm going to give him a little bit of a party. That's all I'm going to do for him. But son, everything I have is yours." That son who stayed home was recipient of all of his father's wealth.

The two words that describe God's essence and God's character are love and holiness. How people perceive God is His love. However, that is not the essence of God's character. The essence of God's character is His holiness. God said that if you want to be like Him, you must be holy. *"...Holy...is the Lord...."* (Isaiah 6:3)

Hebrews 12:9, 10, *"Furthermore we have had fathers of our flesh which corrected us, and we gave them reverence: shall we not much rather be in subjection unto the Father of spirits, and live? For they verily for a few days chastened us after their own pleasure; but he for our profit, that we might be partakers of his holiness."* What an earthly father wants to do is not just give the manifestation of his generosity. His gifts are a manifestation of his love. None of those gifts teach a son about his dad's character.

Your dad would like to teach you his character and his holiness. What your dad is, is his character, and he reveals his character through a mode of operation called love. He expresses that by putting a roof over your head, by giving you food for your belly, by putting clothes on your back, and by taking care of your expenses. He pays for your Christian education and provides transportation and loves you. However, when your dad senses that you have an open heart to him, he may say, "Son, let me teach you something. Let me teach you about women. Let me teach you how to be a good man to a woman. Let me teach you about money. Let me give you some ideas." Your dad begins to share his character with you. When your dad shares his character with you, he is teaching you how you can become what he is so you can be to others what he is to you.

That is what God would love to do with us. That's what every real father who loves his son would like his son to become—a chip off the old block with the character that he has. Some of you are terrified because you do not want your son to become like you are. God says if you are the right kind of a father, you want your son to be a taker, not just of your manifestation which is love, but also of your character. You want him to work hard like you work. You want him to get up out of bed like you get up. You want him to be ethical and honest like you are ethical and honest. You want him to be moral and pure like you are moral and pure. The greatest fear a dad has is not that he won't be loved by his son but that he won't be respected by his son. Many sons say, "I've got a great dad. He's kind. He helps me. If I get in trouble, he bails me out. He's a 'Hail! Hail! The gang's all here' kind of a dad." What that dad worries about is, "I love my son, but he is not copying my character. That hurts so much."

Chastening really only works on those who want to become a partaker of God's character, not just be a recipient of dad's affection. Chastening does not always work in every child's life.

CHAPTER VIII

Rules of Chastisement

"For whom the Lord loveth he chasteneth, and scourgeth every son whom he receiveth." (Hebrews 12:6)

1. Chastisement should re-establish the relationship that exists and the role each plays. In other words, proper discipline validates and strengthens the bond. If I am running a race together with my son and my son gets tempted to turn aside, and he starts going aside, while I've still got his heart, I need to get a hold of that heart and chasten him. The chastening is this, "We're connected. Didn't you know that? Don't you know that you and I are running a race together? You and I are a team. Now you are AWOL on me! I can't win that race without you, and you can't win that race without me. We're in a race that goes all the way back to Abel. I don't want you to do something big so I can feel the glory. This is

you and I on a team."

Chastisement always says, "We are going to get closer than we've ever been before. I'm not going to let you get out of the way." If chastisement does not bring you closer to the one chastened, you have failed in your chastisement. If it does not re-establish and strengthen the roles you play, then you are not doing something right. Chastisement must strengthen the cord that was there.

How do you know if the cord is there? When your son turns aside and you try to chasten him and it drives him further away, then the chastening is wrong, or there was no tie there to begin with. The only reason I would not go after my son is if I did not care about him.

The father says, "Son, I want you back."

If that cord is there, it may stretch quite a ways, but proper chastening always gets them back. You may lose your child a little while. He may step out of the way a little while like Samson stepped away with Delilah. He may step out like David stepped out with Bathsheba, but when Nathan pointed his finger in David's face, and said, "Thou art the man," David did not exercise his position as king and say, "You have no right to do that." He acknowledged the truth of Nathan's accusation and said, "I have sinned, and I'm wrong." Nathan said, "And God receives you back."

David now had to deal with some baggage, but he was greatly used after that because the proper role of chastisement is always to reaffirm the relationship.

If you are spanking your child, and it's driving a wedge between you and him, stop spanking. Rebuild the relationship and find what proper, true, scriptural chastisement is. Chastisement involves infinitely more ingredients than just spanking.

2. Chastening is only effective for those who will listen. The person who is most inclined to pursue God is the one who is most likely to get chastened by God. The Bible says that God chastens everybody, but He works the most on those who listen the most and pursue Him the most. The Bible says, *"Draw nigh to God, and He will draw nigh to you...."* (James 4:8) However, if I don't draw nigh to God, He is not going to draw nigh to me; not because He doesn't want to be with me, but He is going to find Himself wasting His time.

Occasionally, someone comes to me and says, "Did you know So-and-so is not doing very well in the church. They're kind of hitting and missing in church. Do you think you should go after him and run him down?"

I say, "If you'll open the door behind you, you'll see a hundred people lined up three flights of steps who are waiting to come see me. Should I leave these people who want help to go find someone who has already said by his actions that he doesn't want my help?"

A long time ago when I was on my bus route, I learned a great truth. It is easy to waste hours and hours and hours begging and pleading with one arrogant, haughty, rebellious kid to come to church and neglect 30 to 50 kids who, with just a little nudge, would come to church. I decided to concentrate my energies on those who want the help and not waste my energies on those who say, "Leave me alone." I got that truth from the prodigal son who said to his father, "I'm leaving you. There is nothing you can do about it. I'm taking off."

What did the dad do? He said, "Here's your loot. Go scoot."

The kid took off, and dad didn't go after him. I have talked to parents for years about this situation when their adult child decides to leave home. When an adult child is living with you, and he starts getting an attitude about your rules, why wrestle or fight with him? If he wants to make life miserable, just say, "I'm not going to work on you. I'd rather work on my sixth-grade boy who wants to listen to me, or the junior high boys with whom I am working. If you want to make life miserable for us, why don't you go get a job, your own apartment, your own vehicle, and buy your own food and clothing? Why should you mooch off me and make life miserable for me?"

A teenage child is another matter. I think you should tell a teenage child, "I'll decide when you leave home."

I try to invest my time in people who want the help. I'd rather give my time to people who show up regularly for church than fret about the people who don't care enough to come. I am not going to scold the ones who didn't come. I just try to invest my time wisely.

God is saying, "I'm going to invest the greatest amount of My time in those who draw near to Me. I love all My kids, but if you don't want Me, I'm not going to waste My time with you. It's not that I wouldn't give you anything. I'll give you everything; I've already proven that."

Romans 8:32 says, *"He that spared not his own Son, but delivered him up for us all, how shall he not with him also freely give us all things?"* God says, "I'll give you anything you need. I proved that when I died for you. However, if you don't want my help, I'm not going to chase you down and grab you and beat you up about it. I chasten all my children, but I will invest the most in those who draw nigh unto me."

Chastening works best for the simple ones. Chastening is when the simple one is drawing away, and the prudent guide is pulling him back, getting him back on track, and making sure he stays in the race. You can't make anybody do anything. I can barely make me do what I'm supposed to do. I have about a dozen men who report to me every week on how they are doing in some spiritual discipline, such as having trouble with smoking or with a bad attitude. One of them met me a week or two ago and said, "Brother Schaap, can't you just make me do right. I come to you every week. Just make me do right!"

I wanted to tell him, "Son, I can't even make me do right, let alone make you do right."

When a child is four years old, we can make him do right. When he is 14, we can try to make him do right. When he is 24, good luck! The bottom line is you have to win the heart of the simple one and hold it near you. When that heart is near you and the simple one begins to wander, that's when chastening is most effective. The Bible says you can whip him a hundred times, but it won't have as much effect as with the fellow who is right next to you. When the simple one starts to wander, the prudent guide hollers at him and says, "Get back in line. Don't you dare!" This firm word of admonition does more good when the simple one is close to you than would a hundred stripes on his back when he has drifted far from you.

3. Chastisement should be an evidence of love, not anger. Hebrews 12:6 says, *"For whom the Lord loveth he chasteneth, and scourgeth every son whom he receiveth."* When you chasten a child, whether it is a Sunday school child, a teenager on a bus route, a school child in a classroom as a teacher or principal, or if you chasten your own child, it should always be an evidence of love. You make a tremendous mistake when you chasten out of anger. The Bible says in Ephesians 4:26, "Be ye angry, and sin not," and you must make sure your anger is under control, or you will sin. If you sin while you are chastening, you will actually undo the good of the chastening.

If the simple one begins to wander and the prudent guide gets angry at the simple one, he can actually drive him to the scorner. The scorner is so shrewd, he doesn't get mad at the simple one. He always accepts him. Anger with chastening is usually a bad combination. When it is time to spank a child, discipline a child, or administer some type of punishment, a wise parent should always take time to cool down until he is thinking very clearly and very settled. The privilege of being an adult is that you are required to be in control of yourself all of the time.

Everyone has seen people who grab a child with

one hand and beat him with the other. Mama is not going to feel better after a while, and the kid is being disciplined ineffectively. Chastening should always be an evidence of love and not of anger. God said that He chastens those whom He loves. God says that He is not angry when He chastens us. David said, *"O Lord, rebuke me not in thy wrath: neither chasten me in thy hot displeasure."* (Psalm 38:1) David also said, "Dear Lord, You have a right to chasten me. I was wrong with Bathsheba and with Uriah, the Hittite. If anyone deserves to get hit and chastened, I do. But God, do me one favor. I know You love me. Please don't chasten me in Your hot displeasure. When Your anger has really peaked, don't hit me because You are going to hit me so hard, you'll drive me away from You, as frail as I am. The sin that made me do the wrong is evidence of how fragile I am. I'm actually not strong enough to take Your anger."

4. Chastisement should be for the child's profit not the parent's pleasure. Hebrews 12:10 says, *"For they verily for a few days chastened us after their own pleasure; but he for our profit, that we might be partakers of his holiness."* The word *pleasure* means "for show or public display." The place to chasten your children is privately at home. Children should not be spanked in public for any reason. That is not the role of chastening. Chastening is not to show off how tough you are or to show that you are the adult and can handle things. Disciplining should be done in the privacy of your home.

Parents, your job is not to show other parents

what a wonderful parent you are. That's not the role of having children. Your job is not to go to the nursery, whack your child on the bottom, yell and say, "You be good, now." You are not impressing anyone. You are not helping your child at all. You are definitely not going to teach a life principle in 30 seconds of screaming. Save all of your chastisement for behind closed doors when you have hours or long periods of time to train your children.

Also, do not chasten because you are angry at your child. If your child does something wrong in the grocery store, and in your anger you yell at the child, it will not do any good. Get a babysitter and go shopping by yourself if you cannot control your child in the grocery store. If you can't afford a babysitter, swap with someone.

Some people shouldn't take their children to a restaurant. It's not an American duty to disrupt everybody else's meal because you want to eat out. When our children were small, there was a day and night difference between them. In the restaurant, our daughter tended to be very good, and she was Miss Social who loved going out, so she was quite well-behaved.

Every time we would go out to eat with our son, just when we would get our meals he would start screaming and would go on and on. I'd say, "Quiet, Kenny. Quiet!" He would kick and scream. Finally I told my wife, "We're not going to go out to eat until Kenny is about 18!"

If your kids don't behave well, you feel like everyone is looking at you in the restaurant, because they are! You feel like everyone is saying, "Why don't they shut that brat up!" because they are. Don't put yourself in a place of temptation where chastening becomes a way to vent how frustrated you are. That does not rear a good child. Take him home and role play, and try to teach him. If you have a little tyke that is just too young to understand, don't try to force the situation. Eat your meals at home.

5. Chastisement should always build respect for authority. Hebrews 12:9 says, *"Furthermore we have had fathers of our flesh which corrected us, and we gave them reverence: shall we not much rather be in subjection unto the Father of spirits, and live?"* You will know if the chastisement of your children is effective by their increasing respect for you and their other authorities. In our home, I did not chasten the children if they showed disrespect to me. I chastened them if they showed disrespect to my wife. She did not chasten them if they did not respect her. She chastened them if they did not respect me.

One day at the table, our daughter Jaclynn got sassy with me. I looked at her and then looked at my wife. My wife looked at me with a questioning look as if to say, "I don't know if she's teasing or just being funny." Jaclynn got smart with me again a little later.

My wife reached across the table and grabbed her. She reprimanded Jaclynn for disrespecting her dad. She said, "Don't ever talk to your dad like that again."

Before we had children, we had agreed that each would defend the sanctity of the position of the other one. That is why it is so important to defend every other authority's position. What happens in child rearing or a classroom situation is we defend our own position, and we belittle the other person's position. A child hears dad criticize mom while defending his position as dad. The child learns right away that position is only for personal glory.

For that same reason, many do not respect the President of the United States. That is why many do not respect their teachers. We have lost respect for authority. That's why many public school classrooms have to be guarded by guard dogs and metal detectors. That's why you have to have policemen walking the halls in public schools. That's why we have to have security guards by our schools because once in a while we have idiots out there that want to get a little carried away. They want to deface property. They want to take their keys and mark up someone's property, or take eggs and throw them at vans and buses and school property, or they want to smash windows and break into school property. It's called disrespect for authority. That disrespect for authority comes because, in that child's life, the authority figure was not established as a lofty position. A child saw a person of authority defending his position and belittling another's position.

Mom, that is why it is so important that you defend Dad more than you defend yourself. And that's why it's important, Dad, that you defend Mom more than yourself. That's why it is important for parents to defend the teacher more than themselves. That is why it's more important for the teacher to defend the principal more than himself. That's why it's important for the bus worker to defend the bus director more than he defends his own position as the bus captain. That is why it is important for the parents to defend the youth director and the youth program more than they defend themselves. That is also why it is important for the youth program to defend the parents rather than defend themselves. All the authorities must defend each other. Let me defend you, and you defend him, and so forth. No one can say, "What about me?"

I love the statement, "There's no limit to how far God can take you if you don't care who gets the credit." The principle is very good. Mom and dad, if you wouldn't worry so much about how you look in the eyes of the children, but instead are worried about how your spouse looks in their eyes, it would help your image a great deal. God could use you more to help build your children.

6. Chastisement should build a personal strength to stand alone against wrongdoers. Hebrews 12:3, 4 says, *"For consider him that endured such contradiction of sinners against himself, lest ye be wearied and faint in your minds. Ye have not resisted unto blood, striving against sin."* As the prudent guide is trying to get the simple one to wisdom, the simple one is being lured by the siren-song of the independent-minded people. The crowd that is drawing in the simple ones is the rock concert crowd, the rock-a-billy music crowd, the drug crowd. Chastening strengthens the simple one so that he can say, "I don't hate you, but I don't need you. I can stand on

my own hind legs." Chastening always strengthens the person to have the strength to stand alone.

This is one of the great secrets to building strong young men in their teenage years to become capable leaders in their adult lives. I personally don't think leaders are born or made. I think leaders are exposed. I don't know how a leader is made. I know that circumstances reveal whether or not someone is a leader.

Have you ever watched a ball game where a team is behind by 20 points, and the clutch player comes through and rallies his team? I was reading a sports magazine which had the following comment about a certain player, "Everybody thought this player was a dud, but under the pressure, he has risen to the occasion and has surprised all of us and has taken his team to great victory." Why did they think he was a dud? A whole bunch of Heisman trophy winners and rookies-of-the-year were going to make some certain team a winner, but they fizzled and became duds. They were traded and traded because they could not come through under pressure. They had all these promising leadership skills—strength, charisma, charm, and poise, but they couldn't come through when the pressure was on.

Leadership always rises to the occasion. I know that leadership is that one magical, indefinable quality that says, "By myself, within me, all alone, I will find a way to win. I know that." Every leader has it. It's a Douglas MacArthur who got criticized because he didn't say, "We shall return."

Leaders don't say, "We shall." Leaders say, "I shall return." The idea behind his declaration was, "I don't know if my government is going to be with me, but somehow or another, I'm going to come back and deliver the Filipino people."

The press attacked him and said, "That was such an arrogant, haughty statement. It should have been, 'We shall return.' "

"We" abandoned him, and that's why he was getting kicked out of Corregidor. "We" left him to his losses, and he got ordered out of the Philippines. He had to abandon his men, and he was harshly criticized for his action. The truth of the matter is, we starved him out of there and forced him into that situation. He said, "I'll go, and I'll find some way to come back." He did go back, and he did deliver the Filipino people.

Leadership somehow has to find the strength to say, "I don't need the crowd." The scorner and the crowd of sinners lead the simple one to foolishness. The reason why the crowd is so powerful with the simple one is because they give him that false sense of security. Take the Chicago gang banger out of his gang and see how he does by himself. The only strength he has is when he is surrounded by a bunch of punks. That is why it is so important, when a boy is starting to get into trouble, that his relationship with the wrong crowd is completely severed, and he is separated from them.

John the Baptist became such a great leader because he was alone in the desert until his showing. He had to learn to stand by himself. In Luke 3:7, he said, *"O generation of vipers."* How do you stick your finger in the Pharisees' faces and say, "Every cotton-pickin' one of you has missed the truth of this Book, and you are wrong. I'm showing you the Messiah, and you're not smart enough to understand it." How can you get a man with enough strength of character like John the Baptist if he needs the security of the crowd?

A leader cannot have crowd dependency. It is very important to understand that chastening always helps a boy stand alone. When he becomes dependent upon other boys, he is not developing his social skills and his leadership skills.

7.Proper chastisement leads to peace and purity; improper or poor chastisement leads to strife and bitterness. Hebrews 12:12-15 says, *"Wherefore lift up the hands which hang down, and the feeble knees; And make straight paths for your feet, lest that which is lame be turned out of the way; but let it rather be healed. Follow peace with all men, and holiness, without which no man shall see the Lord: Looking diligently lest any man fail of the grace of God; lest any root of bitterness springing up trouble you, and thereby many be defiled."* If the simple one is not properly chastened or handled right, and he doesn't respond correctly to chastisement, eventually he will become bitter. The fruit of bad chastening is always bitterness. Every time I deal with a teenager or a young adult who has some bitterness or some moral decline, it is because of improper chastisement.

A young lady came to my office who was giving her mother fits. I said to her mom, "Would you mind if I spend a few minutes alone and ask your daughter a few questions." The daughter sat there with her arms crossed, and she had that rolling-the-eyeball kind of look on her face, and I said, "Could I ask you just a couple of questions? Then I'll let you go."

"I guess," she answered.

I asked, "Have you ever attended a Christian school?"

"No," she answered.

"Okay, you went to public school, right?"

She said, "Yes."

I asked, "Did someone mishandle you there? Were you ever hurt by an adult?"

Her silence was pregnant with truth. I said, "Tell me about it."

"Well, I was in a church in the south," she said, "and I had a male Sunday school teacher. He scolded me about something." The details are not important, but the bottom line is, this teacher mishandled her, threatened her, and had been ugly with her.

"This problem you are having with your mother is because of that person's mistreating you, isn't it? When a person who is an authority figure really mishandles you, and you're not mature enough to deal with it, you get bitter every time," I said.

She said, "Brother Schaap, you are looking at a very bitter girl. I know I am wrong, but I have a root of bitterness in me, and I can't get rid of it. It comes out with my mom and my sisters. I am still so angry at how I was handled at that church. I'm just furious." She started crying.

There's no way we can be perfect 100% of the time, but I want to encourage all who are teachers and parents and everyone who has a child under his authority to be very, very wise about how you handle the chastening of a child. Hebrews 12:15-17 shows the five-step digression that happens when a child is not handled properly. *"Looking diligently lest any man fail of the grace of God; lest any root of bitterness springing up trouble you, and thereby many be defiled; lest there be any fornicator, or profane person, as Esau, who for one morsel of meat sold his birthright. For ye know how that afterward, when he would have inherited the blessing, he was rejected: for he found no place of repentance, though he sought it carefully with tears."*

• Bitterness. Verse 15 says that first there is a root of bitterness that springs up which leads to many people being defiled. *Defiled* means "contaminated." A bitter person is someone who is living a loose life because of making some very foolish mistakes. The line between those who stay on the up-and-up and don't smoke or drink or take drugs and those who do is so thin. The difference is one word: bitterness.

In my 25 years of preaching, I have yet to find anyone who has lived a life of taking drugs, drinking alcohol, smoking cigarettes, and fornicating who can't point to an issue where he is bitter. I am not saying that it is always justified, but it is always bitterness that brings him to decide to cross the line—always! Bitterness is what causes many to be hurt, families to be contaminated, youth departments to be contaminated, and churches to be contaminated. Many are defiled because of the bitterness of one. That bitterness came in from improper chastisement.

Someone received improper chastening. He was handled wrongly, roughly, abusively, out of anger. He was made a fool of, and humiliated in public, instead of being disciplined quietly at home. Someone was ticked off and wanted to show everyone who was in charge. That bitterness festered until the right opportunity came, and that is when the moral violation occurred.

• Defilement. The word *defilement* means "contaminated" or "moral contamination." One is dirty because of hanging around the dirty crowd and from listening to the dirty music of the dirty crowd.

• Fornicator. The word *fornicator* comes from a Greek word which means, "for sale." The word *fornicator* means, "I am wearing a for-sale sign. I'm bitter, and I feel dirty, and I have a for-sale sign that says that I am available. I'm open to the highest bidder."

There are people who send their young people to Hyles-Anderson College and within two days we are expelling these same young people. Somehow the wrong crowd gravitates to each other; they find each other. It is almost like they wear a big sign that says, "For sale. Purchase me." Did you ever wonder how people who come to a college and don't even know each other are linked up within a week? One might think, "Birds of a feather are flocking together." How did that happen? Bitter people are bleeding, and they smell morally like Limburger cheese.

• Profane person. *Profane* means "crossing the threshold." It means a person crosses an imaginary line or opens the door and walks through to a new horizon. The Bible says that when people get bitter, they are opening a door that says, "I'm going to go some place I would never have thought of going, except now I am willing to consider it." It also says, "I'm opening a door, and I'm letting you come into my life. I'm letting things come into my world I would never have thought."

Young people, when you open that door, it never again fully locks airtight. It's like the door of a car that has been in an accident. You can bump it right, you can nail it shut, you can put a latch on it, but the door always has a whistle when the wind comes through because it has been opened in a way that it never should have been opened.

• Rejection. According to verse 17, the rejected one feels abandonment and disapproval. When the simple opens that door and says, "Hello, scorners and fools, I guess you are my crowd," he is also saying, "I never again feel like I deserve the good crowd."

The classic Bible illustration of rejection is Cain. He killed his brother. He got ticked off and opened the "door" that went to the land of Nod. He left the presence of God and went to the land of Nod, which means "wandering." His words were, "I'll never be accepted back there again. I'm a fugitive and a vagabond. I'll never fit in with that crowd again." Those weren't God's words; those were Cain's words.

When Cain got bitter, he was threatened and cursed by God. A bitter person always winds up feeling that he will never be accepted by the right crowd again. That is not the right crowd's fault. It is the fault of the rejected one, but it is also often the fault of the prudent guide who did not chasten the simple one properly and drove him away, rather than leading him to godliness.

CHAPTER IX

Practical Advice
for Chastening

"Now no chastening for the present seemeth to be joyous, but grievous: nevertheless afterward it yieldeth the peaceable fruit of righteousness unto them which are exercised thereby. Wherefore lift up the hands which hang down, and the feeble knees; And make straight paths for your feet, lest that which is lame be turned out of the way; but let it rather be healed. Follow peace with all men, and holiness, without which no man shall see the Lord: Looking diligently lest any man fail of the grace of God; lest any root of bitterness springing up trouble you, and thereby many be defiled."
(Hebrews 12:11-15)

I have several points to share on the subject of practical advice for chastening.

1. Chastening is likened to training for the Olympics. Hebrews 12:11, *"Now no chastening for the present seemeth to be joyous, but grievous: nevertheless afterward it yieldeth the peaceable fruit of righteousness unto them which are exercised thereby."* The words exercised thereby refer to a man training for "the games." The games were called "Isthmian Games," and they were held near Corinth. These were the forerunner of the Olympic games. They were big, prestigious events for which athletes trained and trained. That is a picture of what chastening is.

We get the idea that chastening is one dimension—punishing, scolding, or chastisement. The Bible says chastening is like a coach's preparing an athlete for the big game, the Olympics, or the once-in-a-lifetime event.

The job of a parent is to be a coach who says, "I'm going to take my child with his raw material, and I'm going to get him ready for the race of life." God says that we are all in a race of life, and He will give us many coaches—parents, teachers, Sunday school teachers, bus workers—and a variety of "coaches" who will pass us on from phase to phase and make sure that we are a gold-medal winner in the Olympics of life.

Chastening is pushing someone further to become better. The athlete is the simple one, and the coach is the parent or the prudent one who is taking that simple one and getting him to wisdom. If you push too hard and too long, you will injure the athlete. You will put him out of the race, and he will lose the competition. He can't even participate! If you push too harshly, you can kill the desire to excel. If you push too little, you produce a loser. This wisdom has to be in the coach or the prudent guide.

The simple one is prone to laziness. He doesn't want to get up early and discipline himself. He doesn't want to work at life, but the truth of the matter is, he wants to stand on the platform of life one day and be an Olympic winner. To do so, he needs someone who is smart enough to get him up. The prudent guide is a coach, and he uses chastening to get the simple one.

Chastening is not just spanking. An athlete is not spanked to get him to perform. If you've ever played under a mean coach who's a winner, he has his own way of chastening you which involves cajoling, bribery, rewards, and words. It involves what we call motivation. The prudent guide inspires the simple one, and his job is to get the simple one to wisdom like a coach would get an athlete to the gold-medal stand in the Olympics.

2. Chastening is multi-faceted.

a. Chastening is explaining. It is teaching. It is the lecture. It is the, "Let me tell you what you must do." It is a sermon. If you chasten a simple one, you need to explain what you're going to do.

b. Chastening is an example. You must be the example to show the simple one how to do it. That's true whether you learn how to drive a car, fly an airplane, water ski, fly a kite, or learn anything in life. Someone explains it, and then you have someone do it as an example. That's what you're supposed to be doing in the home. Mom and dad are supposed to be explaining and then living the example of how to have a good marriage.

c. Chastening is exercise. To chasten someone, the prudent guide needs to exercise what the simple one is learning by example and what he is hearing by instructions. In other words, he has to do the task with you.

When I learned to fly, I got to sit in the pilot's seat during my very first flight. I put $5 down and brought in a coupon from a flying magazine which gave me one free introductory flight.

I walked with the instructor to the airplane, and he said, "Sit in the pilot's seat."

I said, "I get to?"

"Yes, I'm going to show you," he said. "I want you to listen to everything I say, and I want you to do everything I do. I want you to put your hands on the control column."

"Okay," and so I did that.

He pointed and said, "Here's your throttle. Put your hands on the throttle." I put my hand on his hand on the throttle. "See that little dial? That's your RPM meter. We are going to bring it up to where it says, '2,500 RPM's.'"

We pushed on the throttle together. He continued, "Put your feet on those pads on the top of the rudder pedal. That's your brake." So we pushed our feet together there on the rudder pedal. In fact, I took off in the airplane just by mimicking him. He had his hand on one control column, and I had my hands on the other. I had my feet on the rudder pedals in front of me, and he had his feet on the rudder pedals in front of him. He had his hands on the throttle, and he had me put my hands on the throttle on top of his, and we flew the plane together. All of the time, he was explaining what he was doing by doing it himself and having me exercise.

"Your muscles have to learn how to do this," he explained. "So one day we're going to learn it." I got so excited! I plopped down $795 of my hard-earned money to buy the course!

He said, "That buys you the very minimum it takes to get a private pilot's license." I got done and had money coming back to me. So, I got to fly. He taught me everything. He said, "I'm going to tell you what it is, and then I'm going to show you what it is, and then we're going to do it together."

d. Chastening is refining. That refining is nothing more than practice or the perfecting of the exercise. I remember when I was flying for the very first time. The instructor was with me, and I took off with his help. The next time I did it all by myself without the instructor touching the controls. Every time after that I took off by myself.

The first time we landed, he did it, but I had my hands on the controls. The next time I did it with his help, and the third time I landed the plane mostly by myself. After that third landing, he never again landed the plane for me. After seven hours and twelve minutes, I landed the plane, and he climbed out and said, "It's all yours, buddy. You're solo."

Was I perfected? No, because the next day he said, "Now we're going to do some precision work in flying." We went up about 3,000 feet, and he said,

"Do you see that intersection down there off the left wing?"

"Yes, sir," I acknowledged.

He said, "We're going to do turns around a point. I want that wing to stay right on that intersection. Don't let it vary."

I was all over the place with that plane, and he said, "You're pretty miserable."

I thought, "I can take off in a plane and land it, but I'm doing a terrible job!" I said to him, "I'm the most lousy pilot in the world."

"No, you're not," he assured me. "You're doing great." I became so good, that almost without looking, I could spin that airplane around to any angle.

Next, we had to do slow flight. We had to make the plane fly just 5 mph above falling out of the sky. I had to learn what a plane acts like, and how it wobbles when it moves. I was perfecting my skill. Yes, I was taking off and landing every time, but I was refining my flying skills.

Later the instructor said, "Now, we're going to fly 200 miles from point A to point B instead of just flying around the airport." He taught me how to plot a route, how to talk on the microphone, and how to get permission from the FAA to fly to another location. The instructor kept refining, refining, and refining.

One day the check pilot, the senior pilot, came to the airport. He was a fearsome man. I had flown with him one time, and he was a professional. He said, "I'm watching. I'm not going to do it with you. I'm just going to watch, and it has to be flawless." For about an hour and a half, he ran me through my paces. He would say, "Look out the window over there," and he would reach over and shut off the power. He would say, "Shut your eyes, and put on this hood so you can't see." I would have to correct the plane under unusual predicaments. By the time I thought I was finished, I was sweating and had clammy palms. I was relieved when he finally said, "Land it." Just as I was making the final preparations for landing, he said, "Emergency! You can't land it there," and we had to take off again. Then he ordered, "Land it on the grass instead of the pavement." He kept thinking up all kinds of emergencies, and I finally landed. He looked at me and said, "You're pretty good."

"Thank you," I said.

"Not that good, but pretty good. You've got your license," he said and signed off.

What a happy day that was for me! I called all of my friends and said, "We're going flying! I've got my license," and we went flying.

The first thing I did in flying an airplane was have the procedure explained to me. Then, someone showed me by example. We then exercised and then refined.

e. Chastening is punishment. The words, "scolded," "scourged," "corrected," or "chastised" could be used. The Bible uses all of these words synonymously. The idea is that someone has done wrong.

I remember getting into the airplane for the first time, and the pilot said, "Jack, in the pouch next to your leg pull out the plastic card." I did, and he said,

"What does it say at the top?"

"Pre-flight checklist," I read.

"Read that, and then I'll show you where everything is. You will do this checklist every time we get in the plane. Do you understand that?"

I said, "Yes, sir, I do."

He said, "Good," and we went through the list in order. We had to check all of the instruments and all of the electrical switches—everything. We had to make sure the rudders and flaps worked. By the fourth or fifth time, I had memorized that whole list. After all, I wanted to be a pilot. I had dreamed of it since I was three years of age, and I wanted to be a good pilot. So I didn't want to just read the list; I wanted to memorize it. However, my instructor insisted, "You do it this way every time."

Shortly thereafter, I jumped in the plane, began flipping switches and moving my hands over the instrument panels. He looked at me and asked, "What are you doing?"

I said, "I'm pre-flighting."

"No, you're not," he said.

"Yes, I am," I replied.

"NO, YOU'RE NOT!" he scolded. "Get out that plastic card right now! You read it every single time, or I'm getting out of here, and you're not flying! Do you understand?! Some day you are going to be a pilot, and you are going to have people inside the plane. Some day people's lives are going to depend upon your ability to know what's going on. Don't you ever trust your puny memory because you might have had an argument with a friend or had a bad day or a slip in your memory. That ink on that paper is better than your memory. Now, get out that paper and read it every time! Don't you ever let me see you doing that by memory again!"

I said, "Yes, sir!" I got out that paper, and I shouted out the first thing and continued down the list.

He looked at me and said, "That's better."

That was chastening. He was scolding me and scourging me, and if I hadn't done it, he would have gotten out, and I wouldn't have gotten my pilot's license.

If the prudent guide chastens a simple one, it does not mean that he bends the simple one over his knee for a spanking. Depending on the age of the simple one and the propriety of the type of chastening needed, it may be the fifth one on the list.

Perhaps, Mom or Dad or Teacher or Bus Captain or Sunday School Teacher, we should get the order right. The order is not to punish first. Far too often we resort to punishment first. There tends to be a punishment mentality because we get exasperated. The Bible says in Hebrews 12:10, that our fathers *"...chastened us after their own pleasure; but he* [God] *for our profit...."* God says the difference is that He knows how to chasten properly. Parents sometimes spank hard when they never sat down and explained, or never lived the example, or never exercised with the child or got in the harness with him, or never refined and perfected. They just jump to number five in the list and spank.

The highest authority of a child is a parent, not the government. However, because people do not want to obey God's Word or are ignorant of God's Word and do not practice God's Word, some beat their children and hurt their children. Someone has to step in and protect that little child. Unfortunately, but practically speaking, I agree with the government. If you are not going to take care of your child and do it the biblical way, that child's life may well be in danger. As a good neighbor, I've got to step in and say, "Your child's life is in danger, and I've got to protect life. That's the tenth commandment."

What happens is that some people abuse their parental power and go way beyond it. As a result, we have some seemingly absurd rules, and we are scared to death to spank our kids nowadays. Why? Someone might turn you in to DCFS.

Let me say this, spanking is a portion of chastening, but chastening is first explaining, then being an example, then exercising, and then refining. Once they have learned a lesson, understand it, know it better, have carefully gone through it with you, are good at it, and then choose not to follow, the scolding comes in.

If you take the word *chastening*, and you punish first, it may be a part of chastening, but you better try all of chastening. You need to instruct the simple one first. You need to say, "Let me teach you what we're going to do, how to do it, and I'm going to do it with you. I'm going to do it, and show you how it's done. We're going to do it together, and I'm going to help you perfect it. We are going to train and train and train. We're running the race of life. When the Olympics come, I won't be there to help you. You're going to be on your own."

When you reach adulthood, you'll be on your own. You'll be rearing your children, but your parents won't be there. You can take their advice, but they can't train you then. It's all over.

So, the prudent one needs to instruct, explain, give examples to, exercise with, and refine his child. If he chooses to go the other way, say, "Wait a minute." That's where your word *chastening* comes in.

We've said that chastening means to train someone for the Olympic games of life. To do that there are five steps: explanation, example, exercise, refinement, and punishment. Those five words are the full meaning of the word "chastening."

3. Punishment is the last resort, not the first resort. Punishment is to be used when the other four ingredients have failed to produce the desired result of character change. If the people who abuse their children would have first explained, led a good example, exercised, and refined, and then carefully and prudently chastened or corrected, you wouldn't have DCFS nowadays. Every rule that we hate or despise or feel is a cumbersome burden around our necks is there because we have lawbreakers in society. The Bible in I Timothy 1:9 says, *"Knowing this, that the law is not made for a righteous man, but for the lawless and disobedient...."* Unfortunately, we are encumbered with millstones around our necks because of lawbreakers. We want to be law keepers, so we need to carefully understand this concept.

4. There should be seasons of training and times of rest. We cannot train all of the time. There's a little

wording in the Greek that will help in the full explanation of Hebrews 12:13, *"And make straight paths for your feet, lest that which is lame be turned out of the way; but let it rather be healed."* The word *straight* means "rising paths followed by resting paths." Perfectly illustrated, it means rising to the next level and standing on that to which you have risen. We need to rise and stand on what we have learned. It's important that we understand that principle.

All training involves those five previously mentioned basic parts broken down. My first lesson in flying wasn't cross-country navigation. My first lesson in flying included lessons like, "Now, this column is called the yoke. This is the mixture knob. This is the throttle. Those are the brakes. Those are wings. Those are rudders. There are your flaps. That's the propeller." I first learned all of the basic flight instructions—the parts of the plane and the function of each.

Later on, would-be pilots learn about flying cross-country in the fog, how to handle icing conditions, and how to handle emergency procedures. Pilots don't learn those procedures the first day. After that first day, I went home and said, "Guess what I learned? I learned what the flaps do, what the rudder does, and how to make a controlled turn. I took the airplane off the ground."

Later on, I learned aerodynamics. I learned why wings lift an airplane, and what a chord in an airplane means. It's different from a chord on a piano or an organ. I learned some basic physics of flight. I didn't learn it all on the first lesson. I learned it in little bites, little steps, little tiny risers, and then I built on top of that.

Every step is built on adding to what was already learned. That's how a child is trained. So many of us are trying to rear 12-year-old four year olds, or we are trying to rear 19-year-old thirteen-year-olds. It will not work that way.

I told my wife one time, "Let our children grow up slowly. Don't push them." Some folks have little three-year-old boys who learn how to shake hands, they quote Scripture and preach. That's wonderful, but how about finishing potty-training them? I'm for manners and courtesy, but we have done some real damage to our generation. A poised and proper lady had just moved from Chicago, and she said, "I don't fit in this neighborhood. Can you help me? I'm out of my element."

I asked, "What do you mean?"

"I enrolled my daughter in fifth grade. There are four girls in her fifth grade class who are expecting babies. I've talked to their mothers. Brother Jack, my little fifth grade girl doesn't even know the facts of life."

"Good," I said. "Keep her naive as long as you possibly can." I wish I could tell you how naive my wife and I were when we got married. In this society, books about the birds and bees are being given to second graders—burn those books! They don't need to know about those things.

I love the illustration from the life of Corrie Ten Boom. She was the daughter of a born-again Dutch family who hid Jews during the early days of World

War II. They were caught. As a result, they were all sent to concentration camps.

Before the war began, Mr. Ten Boom was a clockmaker. Every so often he had to take the train to Amsterdam to get some materials. He often took someone from the family with him. Corrie said something like this, "Every time I went, we had to pack for several days because we went for sometimes as long as a week. One day when I was about 14 or 15 years old, I said, 'Dad, will you teach me about the facts of life?' "

He said something to this effect, "Corrie, when we go out of town, and you pack that big trunk, who picks up that trunk for you?"

"You do, Dad."

"Who carries the trunk for you?"

"You do, Dad."

"Who takes it off the train?"

"Dad, you control the trunk. You pick up the trunk, you carry the trunk, but can I know about the facts of life?"

"Corrie, who picks up the trunk for you?"

"Dad, you carry the trunk!"

He said, "Good. Let's think of the trunk as knowledge. Knowledge that is yours some day to open and partake of, but I carry it now. Let me carry that knowledge for you until it is time for you to open it up."

Mr. Ten Boom was a pretty smart daddy. We have a generation that is way too mature in so many areas, and that's why the simple ones act so immature in other areas. Now, an 18-year-old kid gets married, and his dad says, "Son, do you think we should talk about the facts of life?"

His son says, "Sure, dad, what do you want to know?"

We have a society that is educated way beyond what they should be. The Bible says to be simple concerning those things. We have 12-year-old kids fornicating, talking about pregnancy, and babies. Twelve year olds should be learning how to take out the garbage and giving a dog a bath. I'd rather have 12 year olds busy with their model airplanes. Kids grow up too fast. Do you know why? They watch too much worldly stuff on television.

We have not stair-stepped, and we have not made straight paths for them. We've made paths that bypass certain risers. They need to take one step, and let them absorb that for a while. Then when they're standing good and solid on that, let them go up the next step. By the way, that's why in the seventh grade the only dating young people should do for that year is to attend the Valentine Banquet. They only need one date a year. Let your kids grow up by taking a step and resting, taking a step and resting, and taking a step and resting.

Dr. Ray Young tells a story about his dad in his book, Honor Thy Father. He starts off telling all that his dad goes through on the battlefield, and then he says, "My dad had just turned 20." Theo Young was 19 years old when he fought in most of his battles. He was fighting for his country at the age of 19. His is an amazing story about how sometimes you have to rise to the occasion and grow up.

Let me say, I don't want our 19-year-old boys having to kill anyone. War is not the preferred state of existence. Those are not the kind of things I want your boys to experience. I want them to be tough enough and masculine enough if they're called upon, but that's not what I'm wanting for them. I want 19-year-old boys to be happy about getting their first car or to be excited because they have reached their sophomore year in college.

You first have to crawl before you walk. Experts learned that one reason why children have a hard time reading as a seven year old or eight year old is because they never went through a crawling stage. There is something developmental about the eye/hand coordination of crawling that teaches a child how to read words in a book. Experiments have actually been conducted in having kids in fifth and sixth grade start crawling all over again. Mom and Dad may have made the mistake of pushing them to run before they even started crawling.

Crawling is necessary, not just physically, but also spiritually, emotionally, and with relationships. Slow down the education process; take one riser and stand on that. Then take one more riser, and stand on that. Chastening is the prudent one knowing how fast the simple one should go toward wisdom. Even though you want him to become wise, you cannot rush him. Desire wisdom for him when he is mature enough to handle it.

It is much harder to handle leadership, authority, wisdom, and money than you think. The wisest man in the world said to be wise but to be careful how much you ask for. Wisdom is a huge burden.

It is tough to have a good marriage. That's why marriages fall apart at the rate of 57 percent. It is hard to handle money. That's why the average American is 7 to 18 percent in debt to credit cards. That is why less than 5 percent of Americans retire with enough money for five years of living. Ninety-five percent of Americans need someone else to take care of them when they retire.

CHAPTER X

What Is the Goal in Chastening?

"Now no chastening for the present seemeth to be joyous, but grievous: nevertheless afterward it yieldeth the peaceable fruit of righteousness unto them which are exercised thereby. Wherefore lift up the hands which hang down, and the feeble knees; And make straight paths for your feet, lest that which is lame be turned out of the way; but let it rather be healed. Follow peace with all men, and holiness, without which no man shall see the Lord: Looking diligently lest any man fail of the grace of God; lest any root of bitterness springing up trouble you, and thereby many be defiled." (Hebrews 12:11-15)

What is the goal in chastening? The answer is found in Hebrews 12:14, *"Follow peace with all men, and holiness, without which no man shall see the Lord.*" The goal of chastening is to develop a character worthy of adult responsibilities.

What is my goal as a prudent guide to bring a child to maturity? What is the goal in chastening? Why do I instruct or explain? Why do I live by example? Why do I help a child exercise what I've trained him to do? Why do I refine and perfect him? Why do I correct him, and even punish him sometimes when he goes out of the way? The reason is that I'm trying to develop in him the character worthy of adult responsibilities.

In these verses from Hebrews 12, God is saying that as a prudent guide—a parent, a teacher, or a coach—we should chasten, spank, live by example, instruct and explain, exercise, refine and cultivate because we want to develop in our children or the simple ones the character that says, "I've earned a wife." Earning a wife does not mean you're physically able to sleep with her.

A kid doesn't automatically win a position on a sports team because he wants to play. The coach puts him through drills and tests to see if he's earned the right to play on the team. God says that is what exercise is. That's what training is. It's all to earn a berth on the team.

God says to the simple one, "I want you to train and exercise because I want you to earn a berth in adult responsibility." We need a generation that understands that they have nothing coming to them except the opportunity to earn it. It's time that we have a generation that doesn't walk around saying, "I

deserve it." You don't deserve a giant screen television. You don't deserve a cell phone of your own.
You don't deserve designer clothing. You don't
deserve a car when you're 16 years of age. You don't
deserve your own bedroom. You don't deserve your
own telephone. You don't deserve anything except
food and raiment, and with that you should be con-
tent. If mom and dad give you clothes to wear and a
place to go that has a roof over it and heat in it, God
says that's all you're going to get. You need to earn
everything beyond that, and you had better prove
that you have earned the right to enter the club of
adulthood.

You don't get there by turning 18 or even 21.
Nothing in the Bible says there's a magical age when
you get wisdom. Did you know there's a boy named
Isaac who was 30 years old and called a lad in the
Bible, and there was a 17-year-old boy named
Joseph who was called a man? It had nothing to do
with age! It has to do with what they had earned.
When Joseph was in Potiphar's house, he earned the
right to run Potiphar's household and be called a
man. It had to do with responsibility.

Isaac, who had not done one thing yet, except be
a favored child, was called a lad at 33 years of age.
He was called a man after that experience when he
said to his father, "You may slay me." That's why
later on Jesus refers to that incident when He talks
about Himself in the book of Hebrews. He said, "I,
as a son, learned obedience." How? By yielding to
His Father as Isaac did when Abraham was about to
slay him.

There is a club, Teenager, called adulthood. It's called a husband, wife, mom, and dad club. It means, "I own my own car." "I own my own house." "I pay my own bills." "I work my own job." "I make my own decisions." "I make my own policies, procedures, and rules." "I make the decisions in my life." "I live where I believe I should live, and I do what I believe I should do." It is, "I am my own man," or "I am my own woman, and I have earned the right to be that."

It bothers me that we have made adult manhood the laughingstock of America. Most of the advertisements these days have a woman bailing a man out of trouble. I wasn't reared that way nor trained that way, and I'm not built that way. I need my wife, and she's my partner, but she doesn't run the home.

Years ago, I was inducted into the hall of manhood. Men I admired, like my dad, would say, "Welcome to being a man, son." I remember many days during my growing-up years that my dad would say, "Son, that's not even remotely close to manhood. I don't care how big you are, I'll still whip you, and you'll do it my way!" He scolded me, chastened me, taught me, and gave me examples. He'd say, "You do it like I do! Do you understand that, boy?" It was drilled into my head.

I remember many occasions when I went to my father-in-law's office. I remember when I asked for permission to date his daughter, and he shook my hand, which told me, "You've earned my confidence to date my daughter." That didn't mean that I had earned the right to marry her. When I asked him if I could buy her an engagement ring, he said, "Where are you?"

I said, "Michigan."

He asked, "How soon can you be here?"

I said, "In about two hours."

He said, "Be here in two hours. I've got to talk to you."

Talking? I did all the listening. He reminded me that I hadn't earned the right yet to have her hand in marriage. There was a day when he walked his daughter down the aisle to marry me.

At 16 years of age, a simple one hasn't earned the right to get married. That's why the prudent one has to chasten them, and by chastening them I don't mean saying, "What's the matter with you, you stupid kid!" I don't mean that at all. I mean teach by your example.

Teach the simple one to learn from his Sunday school teacher. He can teach him some things. The simple one has not earned the right to copy him; he is earning the right to watch and be welcomed later into that station called manhood. Adulthood has nothing to do with age. It has to do with being responsible, well-trained, and well-mannered. It has to do with being well-refined in courtesy, character, and deportment. It is carrying yourself like a man. It is walking with a confident step.

People would ask me when I was dating my wife, "How did you get the courage to ask out Cindy Hyles?"

I said, "Well, it was like this, I walked up to her daddy and looked him in the eye, and I said, 'Brother Hyles, my name is Jack Schaap, and I'd like to have the pleasure of dating your daughter.'"

"You did it like that?"

I said, "Yes, I looked him right in the eye, and I shook his hand."

"How did you do that?"

It's something called manhood. A real man is never intimidated by another real man. You might be intimidated by a position. I was not intimidated by my father-in-law's manhood. His manhood received me when I received him as a man. I was 19 years old when I met him. Was I intimidated by his position? Oh, yes! I was nervous. I was meeting, "The man!" I was also a man, and long before that, I had been welcomed to the ranks of manhood by other men that I had looked to as examples. That's why I watch how the young men in our church shake my hand.

Chastening is earning the right and preparing simple ones to have the character that deserves adult responsibilities. It's not a chore to pay bills, it's a privilege. It's an honor. I own my own house. I like that. Nobody can take it from me. If our economy crashes in a depression, I still have a roof over my head. I like having my own clothes. They're mine, and I own them. I like owning everything I have. The nice thing about it is that everything I have is mine. That's the privilege and the station of being a man.

So, what is the goal in chastening? It is to develop the character within the simple one worthy of the adult responsibilities. You simple ones can decide a little bit how quickly you get it. You can't say at the age of 12, "I'm ready to buy a house." No, but you can save the 50 cents an hour that you're making.

When I met with Brother Hyles after I got engaged, he said, "I suppose you're going to have to rent an apartment."

I said, "I don't think so, if you don't mind."

"What would you like to do?"

I said, "I'd like to buy a house."

He said, "Boy, it's going to cost a little bit to buy a house."

I said, "Well, if I look around, I believe I could afford it."

"Son, you'd have to have a $5,000 or $6,000 down payment to buy a house."

I said, "I think I've got it, Brother Hyles."

He said, "How much have you got in the bank?"

I said, "I've got $12,000 in the bank."

"Where did you get $12,000?"

"I've been saving it since I was 11 years old."

"You've been saving money since you were 11?"

"Yes, sir, I worked for a nickel an hour, then ten cents an hour, then a quarter an hour, then fifty cents an hour, then a dollar an hour, then two dollars an hour, then four dollars an hour, and this year I'm making six dollars an hour."

He said, "And you saved that much money?"

I said, "I've got $12,000 in the bank."

"Wow! Let's go house hunting!"

I remember feeling that pride of being accepted by a man who says, "I'm impressed. I like that." I remember thinking, "That was more fun than preaching a good sermon."

I had the pride of having a real man accept me. If you've never had that, you don't know how good it feels. You don't know how good it is to have a boss say, "Boy, I sure am glad I hired you." You don't know how good it is to have a teacher say, "I feel like I've invested in you, and I feel like you're going to carry what I've given. I feel like I've done something right with you." You don't know what it's like to walk across a platform with a diploma that you've earned, knowing you didn't get someone else's grades because you were a cheat and a liar and a thief. It's a wonderful privilege to be welcomed into the army of manhood.

We're rearing a generation of kids who walk around and say, "My name is Jimmy, take all you gimme!" or "What have you done for me lately?" or "You owe me!" or "If you don't do it, I'll sue or I'll shoot or I'll kill!" or "If they don't treat me right, I'll divorce my parents." All of this stupid, silly nonsense really isn't the fault of the children. It's because of an older generation that did not get them ready to be received into that army of adulthood.

When a Jewish boy was 12 years old, he would be declared the heir of his father's wealth. It was a very important event. At that event it would be said about him, "This is my son, and he is my legal heir." That was almost a legal declaration of manhood.

I can't say I have ever met a 12-year-old man yet. I have met some boys who are 12 years old who are on the way to manhood. I'm proud of them. That's why I don't mind their acting their age, but there may come a point when a young boy has to say, "I'm not going to act my age. I'm going to act my character. I want to act my responsibility."

A young man should do his homework and his chores without dad getting after him to do it. Dad's not going to be there to say, "Take care of your wife." Mom's not going to be there to say, "Did you make your bed? Did you brush your teeth and wash behind your ears?" If you're 16, 17, or 18 and still have to have mama do that for you, then you haven't even gotten close to manhood yet. You haven't earned the right for a girlfriend. You haven't earned the right for a car. You haven't earned the right to have your license.

The whole idea of chastening is to get simple ones ready so we can receive them and say, "Welcome to adulthood. Here are the bills. Here are your responsibilities."

People who know why they're being trained, don't say, "I don't want to do that," when the responsibility comes. They say, "Thank you for the privilege." It's an honor to walk among adults and say, "I deserve to be here. I've earned the right, and the reason I've earned the right is because I've had people chasten me until it got to the point where I said, 'I can handle that.' "

That's the goal of chastening. You don't spank your kids because you're sick and tired of them. You don't spank your child because he ticked you off. You don't throw things at them or abuse them in any way because you're fed up with them. That is not chastening. That's the wrath of man and not the righteousness of God.

CHAPTER XI

The Art of Punishment

"House and riches are the inheritance of fathers: and a prudent wife is from the Lord. Slothfulness casteth into a deep sleep; and an idle soul shall suffer hunger. He that keepeth the commandment keepeth his own soul; but he that despiseth his ways shall die. He that hath pity upon the poor lendeth unto the Lord; and that which he hath given will he pay him again. Chasten thy son while there is hope, and let not thy soul spare for his crying. A man of great wrath shall suffer punishment: for if thou deliver him, yet thou must do it again."
(Proverbs 19:14-19)

*P*unishment is the fifth and final element of chastening. When most people hear the word *chastening*, they automatically think of the word *punishment*. As we learned in earlier chapters, five different words make up the meaning of chastening. Each element is equally important and legitimate. First of all, *chastening* means "teaching or explaining." Secondly, *chastening* means "to be an example." To fail to set the example is to fail to chasten. To scold or to correct is only 20% of the actual meaning of chastening. Thirdly, *chastening* means "exercise or training"—the repetitious doing of right. Fourthly, *chastening* is "refining or perfecting."

If there is trouble, then there is the final element of chastening which is punishment. Punishment is the last resort, not the first resort. Punishment is to be used only when the other four ingredients have failed to produce the desired result.

Chastening is coaching. The prudent guide of the simple is a coach that is training a protégée. The coach is to excel in what he is teaching his students so that the students will look at the authority and say, "I want to be like you." The football coach doesn't go up to his team and say, "Fellows, I'm not very good at football, but let me teach you how to win a Super Bowl." No. The coach is to be someone who excels.

The goal in chastening is to develop a character worthy of adult responsibilities. Parents have failed in their responsibility when their child says, "I don't want to get married." Something is very wrong there. The greatest compliment I have ever received was from my daughter. My family and I were sitting outside at our picnic table enjoying a meal and time together. My daughter, who was engaged at the time, said, "You know what my life-long goal is?"

I said, "What's that?"

She said, "I want to have a marriage with Todd just like you and Mom have." I wanted to jump up and down. I was thrilled. I walked into the house later and said, "Lord, if I could ever make You feel as good as Jaclynn made me feel, I would be the happiest Christian in all the world." To have a child say, "What I want to do and what I want to be is what I have seen in you," is the ultimate reward.

When I ask a high school senior what he wants to do when he graduates, I like it when he says, "I want to do what my dad does." That may not be the will of God, but I admire a boy who wants to be like his dad. The goal of the authority is to get the child to say, "I look forward to taking on responsibility. I look forward to paying my own mortgage payments. I look forward to getting up every morning and going to work." My dad taught me that work is a great, enjoyable, wonderful thing. I didn't always like work, but I liked being able to work.

There are several steps to the art of punishment.

1. Have a clear plan. Each family should have a "family way" of doing everything. At Hyles-Anderson College, I want us to develop the "Hyles-Anderson way" of doing everything. There is a Hyles-Anderson way to shine shoes, to make a knot in a tie, to clean commodes, to make a bed, to sweep the floor, to do homework, and to read the Bible. When there is a way for each aspect of life, that is a path that people can travel. When there is no "way" of doing things, then each person does that which is right in his own eyes.

When a family does not have a way of doing

things, then each child will have his own independent way of thinking that will frustrate the parents. The Smith family should have a "Smith" way of doing things. The Schaap family should have a "Schaap" way of doing things. If not, the family will be confused. If your family does not have a way of doing things, your children will pick up the neighbor's way of doing things. Parents usually explain how they do things while they are trying to swat the child's bottom as he is running in a circle around the parent. The truth of the matter is, when the parent says to a child, "That's not the way we do it!" he really means, "At this moment I don't have the foggiest how we do it, but I don't want it done the way you did it!" Most punishment says, "I don't want it done that way, but don't ask me how we do it because I don't know."

My church staff has a way of doing things. We have prayer time together every morning. Our staff is required to be there. No matter the inconvenience, each staff member is to be in prayer time. Some of my assistant pastors teach at Hyles-Anderson College and must drive back and forth between the church and the college. However, the First Baptist Church way is that if a man is an assistant pastor at the church, he must be in prayer time at 8:30 a.m. Monday through Friday. Our prayer time consists of sometimes reading a joke, answering questions, taking prayer requests and then prayer—nothing life-changing, but that is our way of doing things.

A school teacher should have specific procedures in the classroom: a way to take attendance, a way to collect papers, a way to pass out papers, a way to take tests, a way to answer questions, and a way to handle each aspect of the classroom. The teacher must have a plan.

If a family, a teacher, a bus worker, and all other authorities do not have a plan, a way of doing things, the authority is setting up for someone else to influence the children under their authority. The reason a child does something in a way that displeases an authority is because that authority did not teach the child the way he wanted it done.

Most children grow up left to themselves. Dad comes home, sits on the couch, watches the news, and falls asleep. Mom stays busy with her life, and children for the most part grow up themselves. There is no plan for the children to follow. They come home from school and watch dad sleep on the couch and listen to mom holler and fuss, and then they all go to bed angry at each other.

Families should have a way of eating their family meals, and a way of having family devotions. I know folks who say, "Families must have their devotions right after supper." That's fine. That is their way. I did not have family devotions after our evening meal. I did not make my family stay after dinner for devotions because I wanted the Bible to be fun to my children. I felt that when I said, "I know you want to go outside and play, but sit down. We have to read the Bible," I was equating reading the Bible with keeping them from something they wanted to do. So, I made the Bible more important. I told my children, "If you get ready for bed and climb in bed, I'll let you stay up half an hour later and read the Bible." For some reason, children like to stay up for any reason, even to read the Bible! I found my children liking the Bible more. That was our way. Each family must have their own way.

Each family needs a way of how they eat, how they joke with each other, how they vacation, how they do their chores, how they get dressed, how they handle family problems—a way of doing everything.

2. Have training times. The big mistake authorities make is yelling. "If I've told you once, I've told you a thousand times, don't you ever do that again!" Authorities should say, "Watch me as I do it. Observe my ways, and then you can follow in my steps." A parent should take his child to the bedroom and show the child how to make the bed. Instead, parents say, "Didn't I tell you to make that bed?!" The child says, "I threw the covers over it." That's the child's way of making a bed.

3. Role play. Parents should have a how-not-to session. Kids like them. When my children were small, I would say, "Let's all pretend that we are at the Smith's house. Let's act like we should not act." The children would run upstairs and grab the toys and throw them down the steps. They would climb all over the couch with their shoes and yell and run into the kitchen and open the refrigerator. (Some mothers will scoff at this idea because their house is a museum. If a couple wants to live in a museum, they should not have children.)

Then we would stop the children and say, "Now we get to do it the right way." Then we would practice the right way to behave at the Smith's house.

Role playing makes a dramatic, visual impression on the child of the difference between right and wrong behavior. If parents will role play several times, when the children go to someone else's house and other children are misbehaving, your own children will look at you and say, "They don't know how to do it right!" The children will stand in judgment of their own peers and say, "They forgot to shut off the wrong way and do it the right way."

In my sermon, "America, America," I use large video screens to show for a few brief seconds some of the so-called "Christian" rock 'n' roll groups. We play some of the music for about 20 to 30 seconds. We then shut down the music and an a cappella group of men come out and sing, "Turn Your Eyes Upon Jesus." The contrast is so obvious to the teenagers. Dozens of teenagers have said to me, "I never saw it so plainly. There is a difference between right and wrong music." That is role playing. That is how you show God's people their transgressions.

Role playing is a wonderful way to chasten a child. Of course, you don't let children do sinful things. Let them act out wrong social behavior or wrong emotional behavior such as pouting and lying on the floor and pitching a fit. Then stop the child and tell them to behave the right way.

When I was a youth director years ago, I had a singing tour group come to our church one time. I was so excited about having them there that I had my teenagers get ready to greet the group as they arrived. When the group got out of their van, the first words out of one member's mouth were, "Any mail for us?" In my heart, I wanted to say, "Jerk! The biggest thing to my teenagers is not your stupid mail! If I ever have anything to do with tour groups, I will teach them the first thing to do when they get out of the van."

When I began working at Hyles-Anderson College and was put in charge of the summer tour groups, I trained those young people how to act when they arrived at a church. I told them that when they get out of the van, they don't run inside and take over the piano. I told them that when arriving at a church and getting out of the van, they should say to the first person they see, "My name is So-and-so, and I'm glad to be here." I told them to prepare 10 miles before they arrived at the church by calling off the name of the pastor, reviewing the name of the church, and getting the names down deep in their hearts so they would not accidentally say, "Boy, it's good to be in Tucumcari, New Mexico," when they are really in Apache Junction, Arizona. Then I told them to pray for the pastor and the church and the meeting. When they arrived they should say, "We were praying for you folks on the way here. We're excited to be here! Anything we can do to help you?" Before you start moving in your PA equipment, you should say, "Pastor, I'm the representative for our tour preacher, and on his behalf may I ask a favor? Could we possibly bring in our PA system and set it up at your convenience? If you don't mind, our pianist would like to get comfortable with your piano. With your permission, may she come in and practice?" We trained our tour group members.

Pastors would write me and say, "I love your singing groups. They are so polite. They don't take over the church when they get here."

Role play in your schools. Role play on your bus routes. Role play in your homes.

4. Post clear boundaries. There must be distinction between right and wrong, with no shades of gray. Never say to your young children, "I'm not sure about that." You might say that to a young adult to guide him to your position, but you do not use that statement with simple-minded children. There must be a clear boundary between right and wrong music, right and wrong attitudes, right and wrong manners—it has to be there.

Children must have an absolute curfew. Each should have a chore list which outlines his responsibilities.

5. Most punishment is actually a failure in training. General rules do not demand specific obedience, and thus cannot require specific punishment. That is provoking a child to wrath because a child does not see the black-and-white of the issue. The child then relates all forms of punishment with how the authority feels that day.

Some parents who are extremely strict are not able to rear children to that level of extreme strictness. Thus, they are building in their children an automatic feeling of despising righteousness. Choose a level of standard that can be lived consistently seven days a week.

Mothers come to me and say, "My husband gets on his kicks. Sometimes he will sleep in until 9:00

a.m., and then the next morning he yells at us because we are not up at 5:00 a.m. reading the Bible." Parents, if you don't get up consistently at 5:00 a.m., then don't yell at your children because they don't either. Some children don't know what to follow—a parent's example or his mouth. Find a time where you can consistently say, "At this time, our family gets out of bed." The time is not the issue; the consistent level of behavior is the issue.

A man came to me and said, "I told my kids, 'You're in high school. I don't care how old you are. I'm the man of the house. Curfew is 9:00 p.m.' "

"Okay," I said, "but we have a problem here."

He said, "Tell me about it. My kids don't obey me."

I asked, "Do you know what time our basketball games end? I didn't leave the basketball games when my son played until between 9:00 or 9:30 p.m. We've already broken your curfew rules."

"I'm the man of the house," he said.

"Yes, you are the man of the house. However, you are telling your children that you are stricter than everyone else is. You may do that, but you are breeding frustration in your children. Why don't we come up with a compromise? Why don't you choose 9:45 or 10:00?"

"You let your kids stay out that late?"

"Oh, yes," I said. "Even worse, I stay out that late."

Parents must find a medium that they can consistently hit. When my children were in activities, I looked at the program schedule and found a time that my children could consistently come home on time. So whether they were in the youth leagues or participated in any of the church activities, they could still come home by curfew and be obedient.

6. Establish specific punishment. There must be a compensatory punishment for the infraction. Not every violation is worthy of capital punishment. The specific punishment should be different for the age group. A teenager may willingly take five swats because he broke curfew with his girlfriend, whereas a ten-year-old boy who did something wrong might greatly dread the five swats. An equal punishment for the teenager would be for the parent to say, "You were out past curfew with your girlfriend, so for the next 30 days you won't see her at all." The authority must find a judgment that is just compensation so that the child will say, "I don't ever want to do that again." We had seven reasons established for which we would spank our children. It was automatic because they broke the rule, not because I was angry. Punishment must be defined and set. When the judge says, "That's 90 days in jail and a $1,000 fine," when the 90 days are over the judge doesn't say, "I'm still angry. Ninety more days!" The authority cannot keep going on with the punishment. Punishment must have a beginning and an ending.

7. It is best to have a place to punish. Generally, the place to eat is in the kitchen. Generally, most people sleep in the bedroom. Generally, an occupation is confined at the workplace. There should be a place where the authority says, "Go to...." The place doesn't matter. I don't know why, but I always sent

my children to the bathroom. Whatever room is chosen, the place should be associated with punishment. Don't take the child there or drag him there—send him there.

8. Never punish when you are in the heat of anger. This accomplishes nothing except breeding contempt in the child. The child should be waiting for you for a while. Many parents punish because they get ticked, not because the child broke a rule. While the child is waiting, the parent should calm down and then administer the punishment. The parent must prepare and walk in with the proper posture and gravity of face.

9. The leader must accept the responsibility and the blame. The leader must say, "We're here today because it is my fault. I am your example. I am your father, and I have failed you. Punishment is the last item on my list of trying to teach you how to be a responsible adult. Evidently, I have failed to train you. I've tried to teach you what is right. I've tried to show you by example. I've tried to exercise with you how to do it right. I think I've told you how to do it, but I must have failed somehow, because if I was as good a father as I should be, I don't think you would have done what you did." The leader is not saying that he bears the guilt, but he does accept the responsibility and the blame.

10. The simple one must admit the guilt. The leader does not bear the guilt if he did not commit the act. The leader takes responsibility, but the simple one must bear the guilt. When Jesus said in Luke 23:34, *"Father, forgive them; for they know not what*

they do," that did not mean that He would not send those to Hell that did not get saved. The guilty must admit his guilt.

CHAPTER XII

How Do You Chasten a Scorner?

"Wherefore hear the word of the Lord, ye scornful men, that rule this people which is in Jerusalem. Because ye have said, We have made a covenant with death, and with hell are we at agreement; when the overflowing scourge shall pass through, it shall not come unto us: for we have made lies our refuge, and under falsehood have we hid ourselves." (Isaiah 28:14, 15)

*T*he chastisement for a scorner is radically different from the chastisement of a simple one. If the prudent guide chastens a scorner with the methods for a simple one, the prudent man will become a fool in the scorner's eyes. If the prudent guide chastens a simple one with the methods for a scorner, the simple one will be driven farther away from the prudent guide; his spirit will be destroyed.

There are many talented people who will never reach their potential because someone treated the simple one with the chastisement for the scorner. There are scorners who have been treated like simple ones who have mocked the prudent guide, and the prudent man wonders what to do with that scorner.

What is the proper way to correct a scorner? Isaiah 28:14, 15 describes five characteristics of a scorner. In those descriptions a method of how the scorner should be disciplined is given.

1. A scorner desires control. *"...Ye scornful men, that rule this people which is in Jerusalem."* Scorners try to seize control or have a position of authority. At the base of all scorning is a desire to control.

2. A scorner is bound by a covenant, or they would rather die than change. *"We have made a covenant with death...."* No matter if authority scolds him or spanks him, a scorner will not change. A scorner is so fixed on where he is and what he believes, he does not care what the teacher or the pastor or parent says to him. He believes he knows what is best for him and will do what he believes is right for him and would rather die than change.

3. A scorner finds his pleasure in the kingdom of darkness. *"...with hell are we at agreement."* The scorner says, "I don't care if you tell me wisdom will bring me money or happiness or success, I refuse to go toward wisdom because I have bought into the kingdom of darkness. I don't want your 'Amazing Grace.' I want my music, and nobody is going to tell

me that I have to give up my rock 'n' roll. Nobody is going to tell me anything."

A scorner can hear the Gospel preached, the truth preached, and he will say, "I don't care what you say. I am not going to change." Authority cannot logic with a scorner about the wrongs of marijuana or alcohol. A scorner is going to do what he is going to do.

4. A scorner has a false sense of security. *"...when the overflowing scourge shall pass through, it shall not come unto us: for we have made lies our refuge."* A scorner can be told, "Cigarette smoking causes cancer," and he will say, "Prove it!" A scorner can be taken to the cancer ward and shown the effects of smoking, and he will say, "It's not going to happen to me." He has a false sense of security that says, "I won't get caught." When the scorner does get caught, he says, "I'll get out of this." The scorner really believes there is no God or judgment.

5. A scorner is a liar. *"...and under falsehood have we hid ourselves."* Scorners stick together with their buddies and continually lie.

How to Discipline a Scorner

What is the proper way to discipline a scorner? What does the principal do when he has a scorner in his school? What does the Sunday school teacher do when he has a scorner in his classroom?

1. Do not give the scorner any authority. It is a mistake for a prudent guide to make a scorner a leader in order to give the scorner what he wants or to try to win the heart of the scorner. The last thing authority should ever do is reward a crook by giving

him control of the bank. Strip the scorner of all authority. A scorner who is given a little authority or a little room to negotiate will contaminate an entire group.

Never use position to try to gain advantage on a scorner. It will always backfire. A coach should never say, "I'll put you on the team" to try to help a scorner. If a team has a scorner, the best response from the coach would be to bench that boy and kick him off the team. Everyone has had to go through the ordeal of watching a scorner give his coach and the referees grief. The scorner wants to question every call and every move.

2. Break up cliques of scorners. There are good cliques and bad cliques. Peter, James, and John were a good clique. I approve of good cliques. Wise people and simple people need to have cliques. Proverbs 13:20a says, *"He that walketh with wise men shall be wise."* I have no problem with my children having tight friendships. That kind of clique is a way to ensure children have a solid "iron fence" around them to make sure the entire group is going together.

Sanballat, Tobiah, and Geshem formed a bad clique that gave Nehemiah fits. They wrote critical letters to Nehemiah and tried everything they could to stop him. Dathan and Abiram were part of a bad clique of scorners who accused Moses of taking too much authority. God allowed the ground to swallow up those scorners. At the foot of the cross, some people scorned Jesus and said, *"He saved others; himself he cannot save."* (Matthew 27:42a) Job's three friends were a clique of scorners. Encourage the friendships of the wise; dismantle the friendships of the scorners.

3. Maintain strict standards and strong preaching. The temptation for authority is to be cool and try to win the heart of the scorner. Authority is never "cool" enough for the scorner. The best thing to do is stand and say, "Thus saith the Lord!" That is what Moses and Joshua and Nehemiah did. The leaders all said, "I'm not going to put up with you. There's the line. Cross it or go to Hell!" Teenagers do not need "cool" leaders. Teenagers need righteous and holy leaders. It is not the authority's job to condescend to the level of the teenager; it is the teenager's job to grow up to become like the authority. Decent teenagers want hard preaching. They want a cause and a high standard. They don't need some "cool dude" playing his Gospel rock. Over 7,000 teenagers come to Youth Conference at First Baptist Church of Hammond each year to hear that kind of strong preaching.

4. Be very consistent. Scorners are looking for a "chink" in authority's armor. The scorner is looking for the leader's inconsistencies in what he preaches and how he lives. To find the inconsistencies of the parents, teachers, Sunday school teachers, and preachers are exactly why scorners exist. The scorner says, "I have heard your lips flap, and I've seen your life lived. What you preach and what you live are far apart. You can have your religion!" Dr. Lee Roberson said the number-one ingredient of a leader must be consistency.

Dad, do your children see you walk with God?

Mama, do your children see you dust off the Bible and find a quick lesson for your Sunday school class as you hurry out the door? Do they hear you teach your Sunday school kids about the unsearchable riches of the Bible while they know you watch Hollywood movies and soap operas and talk television and hardly ever open the Bible? Do your children see you live an inconsistent life? The scorner looks at the inconsistency and hypocrisy in the lives of his leaders and says, "I don't want that hypocrisy any more." Inconsistent leadership is a security of scorners.

5. Demand the truth, and live by the truth. The one thing I will not accept from our students is a lie. I will be gracious and helpful with students who commit a variety of infractions. However, a student better never lie to me. Authority must never let lying go unpunished. God cannot countenance a liar. He said in Revelation 21:8 that all liars shall have their part in the lake of fire. Scorners are liars. The scorners who mocked Jesus at Calvary and made a fool of Moses and defied Nehemiah and ridiculed Job were all liars. The single most important thing to teach a child is the truth.

From day one that I became pastor of the First Baptist Church, I told my staff, "Here's my ministry and the platform on which we stand: We tell the truth to everybody." I told my deacons, "I have no experience to give you, but I have truth to tell you." The strength of my predecessor, Brother Hyles, was his integrity. He told the truth.

The scorner lives in a fabricated world. The truth

is his greatest enemy because the truth exposes the darkness of his heart. The prudent guide will say, "We stand for the truth and righteousness. We tell the truth. We speak the truth. We live the truth." The harshest punishments should be for lying.

These five chastening techniques don't say much about reclaiming the scorner. I have studied the Bible from Genesis to Revelation, and there is not one incident of a scorner being reclaimed. There is more hope of a fool being reclaimed than of a scorner. A fool could be an ignorant, untrained, untaught, mislead, misguided simple one who is just caught up in the foolishness. A scorner is a planning, conniving deceiver who is as shrewd and as clever as Satan. In the Bible, Satan never says, "I repent." Even after 1,000 years in the bottomless pit, Satan comes out with an attitude. He doesn't come out of the smoke of Hell and say, "God, You were right, and I'm wrong. Give me a second chance, and I will bow at Your knees and cry 'Glory, Thou art the Lord' and do anything You want." Instead, Satan comes out with a vengeance, and he is so angry and so clever that he gathers an innumerable army on the fields of the plains of Russia and has a final battle called the battle of Gog and Magog—his one final defiant, "I told you so, God!" hour.

Joseph Stalin's biography, which was written by his daughter, says that Stalin was being trained in the ministry as a teenage boy. However, the inconsistency and hypocrisy of the priests who taught Stalin made him reject Christianity, God, and the Bible. Stalin became a defiant scorner. On his deathbed, he took his final breath and his most trusted advisor said, "Thank God, he is gone."

In one final act, Stalin lunged up, shook his fist toward Heaven, blasphemed God's name, and died.

The prudent guide must grab the heart of the simple one when he is a small child. Parents, teachers, coaches, Sunday school teachers, pastors, and all authorities must do whatever they have to do to get the heart of the child and bring that child to wisdom.

Printed in the United States
24905LVS00001B/28-69

9 781594 675942